SOCIOECONOMIC IMPACT ASSESSMENT OF THE MALALISON ISLAND SOLAR PHOTOVOLTAIC HYBRID PROJECT IN THE PHILIPPINES

OCTOBER 2024

ASIAN DEVELOPMENT BANK

6 ADB Avenue, Mandaluyong City, 1550 Metro Manila, Philippines
Tel +63 2 8632 4444; Fax +63 2 8636 2444
www.adb.org

ISBN 978-92-9270-915-0 (print); 978-92-9270-916-7 (PDF); 978-92-9270-917-4 (ebook)
Publication Stock No. TCS240459-2
DOI: http://dx.doi.org/10.22617/TCS240459-2

Notes:
In this publication, "₱" refers to Philippine pesos and "$" refers to United States dollars.
ADB recognizes "Korea" as the Republic of Korea.

Cover design by Paolo B. Almazar.

On the cover: Malalison Island solar photovoltaic hybrid power plant consisting of a 50-kilowatt photovoltaic system with 273-kilowatt-hour lithium-ion batteries and a 54-kilowatt diesel back-up generator designed to produce 200 kilowatts power, 24 hours, 7 days a weeks (photo sourced from the documentation during the inauguration of the Malalison Island Solar Photovoltaic Hybrid Project in the Philippines).

CONTENTS

TABLES, FIGURES, AND BOX

TABLES

FIGURES

BOX

ACKNOWLEDGMENTS

This *Socioeconomic Assessment Report of the Malalison Island Solar Photovoltaic Hybrid Project* is an output of the Energy Sector Office, Sectors Group of the Asian Development Bank under its regional technical assistance 9960: Pilot-Testing of Innovative Energy Technologies and Business Models (Subproject 3).

The study was conducted by a team in the Energy Sector Office led by Kee-Yung Nam, principal energy economist, with the guidance of Priyantha Wijayatunga, senior director, Energy Sector Office, Sectors Group. The team comprised Grace Yeneza, Felicisima Arriola, and Lyndree Malang, consultants. The management and staff of the Antique Electric Cooperative and its private sector partner, One Renewable Energy Enterprise, Inc., also contributed important data and insights. A group of international and national experts provided invaluable contributions as peer reviewers. Charity Torregosa, senior energy officer; Maria Dona Aliboso, operations analyst; and Marinette de Jesus Glo, operations assistant provided technical advisory and administrative support. Ma. Theresa Mercado copyedited the report, and Alvin Tubio did the layout. The study also benefited from insights and comments of Asian Development Bank colleagues from the Energy Sector Office, Sectors Group.

The publication of this socioeconomic assessment report is funded by the Asian Clean Energy Fund and Clean Energy Fund under the Clean Energy Financing Partnership Facility.

ABBREVIATIONS

ADB	Asian Development Bank
ANTECO	Antique Electric Cooperative
CO_2	carbon dioxide
COVID-19	coronavirus disease
DMC	developing member country
DRES	distributed renewable energy system
EC-PSP	electric cooperatives in partnership with the private sector
EMS	energy management system
ESS	energy storage system
GHG	greenhouse gas
HOMER	Hybrid Optimization of Multiple Energy Resources
LGU	local government unit
NEA	National Electrification Administration
NPC-SPUG	National Power Corporation – Small Power Utilities Group
OREEi	One Renewable Energy Enterprises, Inc.
PV	photovoltaic
STS	synchronous transport signal
UCME	Universal Charge for Missionary Electrification

WEIGHTS AND MEASURES

24/7	24 hours, 7 days a week
kW	kilowatt
kWh	kilowatt-hour
m^2	square meter
mm	millimeter

EXECUTIVE SUMMARY

The Philippines' pursuit for total electrification has been a challenge for over 6 decades. Efforts have been constrained by the country's archipelagic characteristics—comprising 7,641 islands, with many smaller islands beyond the reach of electricity grids. While the country's electrification status reached 91.1% as of June 2023, around 6.13 million households (based on the 2020 population count) remain unserved, with an estimated 1.29 million households located in off-grid areas.

Due to limited or non-access to electricity, development in the country remains unequal. Economic progress in many rural areas is hampered by the lack of reliable affordable and sufficient electricity supply. The National Power Corporation-Small Power Utilities Group (NPC-SPUG) generates power for small islands and remote communities using diesel-generating sets. But many of these areas are supplied with only 4–12 hours of electricity per day due to low population density, minimal demand, and high cost of diesel fuel. Even so, the Government of the Philippines subsidizes NPC-SPUG's operation to keep tariff rates affordable. Continuing in this pathway is not sustainable. Off the grid, 91% of power generation relies on fossil fuel, adding to the country's vulnerability to climate change.

Electric cooperatives have the mandate to provide total electricity access within their franchise area. In areas that NPC-SPUG is unable to serve, electric cooperatives are left to cope with the burden of power generation and electricity distribution, particularly within their franchise coverage. The Antique Electric Cooperative (ANTECO) is faced with this problem in the case of Malalison Island. With its limited resource, it was able to operate a 25-kilowatt (kW) diesel-generating set in the island for 4 hours daily. But with growing demand, ANTECO needed to find an innovative technology and management solution to provide sufficient and reliable 24/7 electricity service in the island and improve its operational efficiencies.

Globally, the use of distributed renewable energy systems (DRES) to generate electricity in areas not connected to the main grids has been increasing. DRES solutions—using such renewable energy resources as solar photovoltaic (PV), bio-power, wind, and hydropower—can facilitate energy access, promote economic development, help curb the worsening climate change, and improve climate resilience.

Hybridizing the diesel-power generation of Malalison Island was deemed as the solution that ANTECO could take.

The support of the Asian Development Bank (ADB) paved the way for ANTECO to implement the solar PV hybrid technology in Malalison Island, Antique Province in partnership with the National Electrification Administration (NEA). The Malalison Island Solar Photovoltaic Hybrid Project features a solar PV hybrid power plant consisting of a 50-kW PV system with 273-kWh lithium-ion batteries and a 54-kW diesel back-up generator designed to produce 200 kW power, 24/7. The project also includes a prepayment metering system, achieving inclusivity, improving ANTECO's operational efficiency, and ensuring 100% collection efficiency.

The project is envisioned to move renewable energy hybrid projects toward commerciality through the partnership of electric cooperatives and the private sector, with a view to potentially attract private sector investment for upscaling the NEA electrification program. A private sector investor partner, One Renewable Energy Enterprises, Inc, (OREEi) was selected through a competitive process. ADB's support of this project is in line with its objective of helping its developing member countries like the Philippines find ways to extend sufficient and reliable energy access services to areas beyond the reach of existing electricity grids.

Assessing the impact of energy projects, particularly at the end-user level is the next important step to take after a few years of project completion. This process determines whether the project has successfully addressed the condition it intended to resolve. However, there is no straightforward method or standard for assessing the impact of rural electrification projects. But what is important to note is that electrification is not merely a household connection or an electric pole in a village or an electric bulb in a house. Energy access brings about multifaceted benefits that impacts people's lives, and these must be recognized. In assessing the impact of a project, the examination should therefore look into the changes that the project brought in the lives of the people and the community. The assessment must consider not only the socioeconomic impact but also other factors such as environmental sustainability as well as energy security.

Using this broader perspective, the socioeconomic impact of the Malalison Island Solar PV Hybrid Project is viewed against the backdrop of the energy trilemma: energy security, energy equity, and environmental sustainability using a set of indicators, which includes a social dimension, such as social acceptance, improved social services, and livelihood enhancements that are also requisite to sustainability.

The conditions at the households and the community before and after the project's implementation are assessed to determine the changes that have occurred. The measurements and values representing the impact relative to each of the indicators are generated from the results of household surveys that were conducted before and after the project, as well as operational data and reports from ANTECO.

The assessment results confirmed that the Malalison Island Solar PV Hybrid Project achieved positive outcomes in all fronts of the energy trilemma.

The project was able to address the energy access and energy equity issues in the island. It enhanced the power generation capacity and extended the hours of ANTECO's service from 4 hours to 24 hours daily. Average monthly consumption per household increased largely due to the availability of 24/7 electricity service, which allowed the utilization of additional appliances, longer hours of commercial operations, and the lowered electricity tariff. The prepayment metering system enabled households to purchase their electricity according to their affordability and available cashflow. Environmentally, the project was estimated to have mitigated greenhouse gas emissions corresponding to 381.48 tons of carbon dioxide (CO_2) equivalent or 84.77 tons CO_2 equivalent annually over the last 4.5 years. Seventy-seven percent of the power generated by the project was sourced from solar energy. Moreover, the residents, especially women and children, were freed from exposure to kerosene lamp fumes and indoor pollution from cooking with fuelwood. With the project, basic services and the general living standards improved. The people were empowered to engage in their own businesses and work for a better future. These positive interventions were made possible through the partnership efforts of the major stakeholders of the project: ANTECO and its private sector partner, OREEi, the local government unit of Culasi and barangay unit of Malalison, and the people of Malalison Island with the support of ADB and NEA.

The project highlights the value of partnerships and collaboration in the introduction of innovative and strategic solutions to achieve shared development goals. Its implementation validates ADB's initial findings that hybridizing the diesel-based power supply generation in small islands in the Philippines is a valid technical solution to facilitate off-grid electrification. It also affirms that there is private sector interest in off-grid electrification and given the opportunity, the private sector can share and contribute valuable knowledge and management skills to improve and ensure viability of projects. Local governments also have substantive roles in making these projects happen.

The report concludes that projects like the Malalison Island Solar PV Hybrid Project should be replicated in other parts of the country and pursued with heightened vigor. This strategy would benefit the remaining over 1,200 island-barangays and about 1.29 million households living in off-grid areas without access to sufficient, reliable, and clean energy. It will help achieve the government's goal of 35% renewable energy in the country's energy mix by 2030 and 50% by 2040, as embodied in the updated National Renewable Energy Program. To scale up implementation of projects from pilot phase to full-scale programs, the financing support of development organizations, such as ADB, will catalyze the achievement of the all-important vision of energy for all with no one left behind.

1 INTRODUCTION

Total electrification has been the policy of the Government of the Philippines as early as the 1960s. However, realization of this goal is constrained largely by the country's archipelagic characteristic, comprising 7,641 islands. At the current scenario, and with population growing, the task remains formidable. Based on the Philippine National Electrification Roadmap, as of June 2023, the country's household electrification level has reached 91.1%, with still an estimated 6.13 million unserved households needing access by 2028 (Census on Population and Housing 2020). i.e., 2.45 million households by end 2023 and 3.68 million households between 2024–2028. Of these households, around 1.29 million are in the off-grid areas.[1]

The National Power Corporation through its Small Power Utilities Group (NPC-SPUG) generates power for small island mini-grids using diesel-fueled generating sets. However, low population density, minimal market demand, plus the high cost of delivering diesel fuel to these islands cannot economically justify 24-hour operations. Thus, in many of these areas, electricity service is limited to only about 4–12 hours per day. At the time of project preparation, 59% of NPC-SPUG areas have less than 8 hours of electricity daily and only 12% receive 24 hours of electricity. Substantial government subsidies had to be provided to maintain these off-grid operations. Moreover, the use of diesel-fueled generating sets, while easy to deploy, lends to unreliability due to difficulties in transporting fuel to the islands, especially during typhoons and monsoons that are prevalent in many parts of the country. And, wherever NPC-SPUG is unable to serve, the electric cooperatives are often left to cope and share the burden of both power generation and distribution of electricity for remote communities within their franchise.

On top of its energy access challenges, the country must also contend with its vulnerability to climate change. According to the World Risk Report 2022, the Philippines (WRI 46.82) is top ranked among countries with the highest disaster risk worldwide with India (WRI 42.31), and Indonesia (WRI 41.46) coming second and third, respectively.[2] Consequently, it is incumbent upon the government to make every effort to transition from fossil-based fuels to more environment-friendly and resilient energy source to protect its people from climate risks. In 2008, the Philippine Congress enacted the Renewable Energy Act to prioritize the utilization of the country's vast renewable energy resources. However, while the share of renewable energy in the country's energy mix at that time was already 33.9%, this share has been declining since then. Instead, the country continues to depend on fossil fuel both on grid and off-grid.

1 Government of the Philippines, Department of Energy. 2023. *2023–2032 National Total Electrification Roadmap.*

2 Bündnis Entwicklung Hilft Ruhr University Bochum, Institute for International Law of Peace and Armed Conflict (IFHV). 2022. *World Risk Report 2022.* p. 6. Retrieved from https://weltrisikobericht.de/wp-content/uploads/2022/09/WorldRiskReport-2022_Online.pdf.

As of June 2022, coal is found to be dominant at 43.1% on the grid, while oil contributes 14.5%, and natural gas 12.8%. Only 29.5% of the energy capacity mix on the grid comes from renewable energy. Off the grid, 91% of power is fossil based with oil contributing 89%, as of 2020 (Figure 1). The country sorely needs to reverse this trend if it were to deliver on its nationally determined contribution commitment to the Paris Agreement and reduce its own vulnerability to climate risks. For this reason, the government has updated its National Renewable Energy Program 2020–2040 with a set target of 35% renewable energy generation by 2030 and 50% by 2040.[3]

Figure 1: Philippine Power Sources

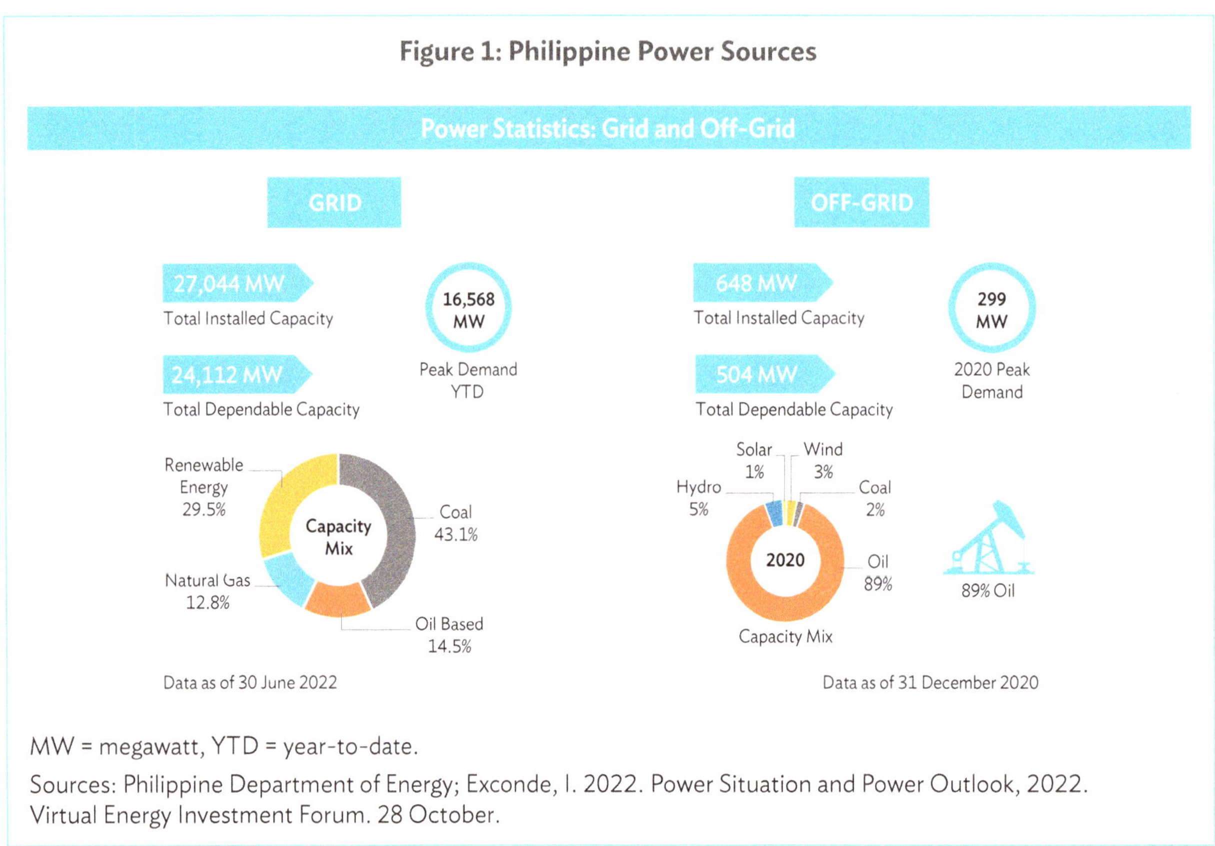

MW = megawatt, YTD = year-to-date.

Sources: Philippine Department of Energy; Exconde, I. 2022. Power Situation and Power Outlook, 2022. Virtual Energy Investment Forum. 28 October.

In response to the government's quest for a suitable clean energy solution, the Asian Development Bank (ADB), supported the implementation of the pilot Malalison Island Solar Photovoltaic (PV) Hybrid Project. The project is in line with ADB's objective of helping its developing member countries find ways to extend sufficient and reliable energy access services to areas beyond the reach of existing electricity grids. This assistance was granted through ADB's regional technical assistance CDTA-0017 REG: Promoting Sustainable Energy for All in Asia and the Pacific – Sustainable Energy for All Regional Hub for Asia and the Pacific – Subproject D: Project Development and Investment Facilitation. As conceived, the project will apply the solar PV hybrid technology in Malalison Island, Antique Province in partnership with the National Electrification Administration (NEA) and Antique Electric Cooperative (ANTECO).

3 Government of the Philippines, Department of Energy, National Renewable Energy Board. 2022. *National Renewable Energy Program 2020–2040*. Renewable Energy Management Bureau Department of Energy.

This type of renewable energy hybrid system was earlier demonstrated by ADB, the Korea Energy Agency, and NEA in Cobrador Island, Romblon, within the franchise of Romblon Electric Cooperative. To move toward commerciality, this pilot initiative will be pursued using the electric cooperatives in partnership with the private sector (EC-PSP) approach, with the view to potentially attracting private sector investment for upscaling the NEA electrification program.

This report examines the outcome of the pilot solar PV hybrid project on the lives of the people of Malalison Island and the changes it brought into the community. It assesses the socioeconomic impact of the project and answers the following questions, among others: What changes have occurred after the solar PV hybrid project was implemented? Did the project improve the power supply situation in the island? Did it create positive impact to the community and to the individual households? Are there any operational issues that ANTECO should attend to? What lessons can be learned from the experience of ANTECO and the people of Malalison Island?

Chapter 2 presents the objectives of the pilot project and relates how the site was selected and the studies that were done to determine the choice of technology and its design, the business model, and use of prepayment system as a management tool to improve operational efficiency.

Chapter 3 discusses the assessment framework and parameters used to assess the socioeconomic impact of the project, both from the perspective of the households and the community.

Chapter 4 shares the findings on the socioeconomic impact of the project based on the household survey conducted in 2021 and data from project operations supplied by ANTECO.

Chapter 5 provides a summary of the assessment findings from the viewpoint of the energy trilemma: energy security, energy equity, and environmental sustainability as well as the outcomes from the social welfare dimension, focusing on the impact to households and the community.

Chapter 6 discusses the replication potential of the project.

Chapter 7 presents some conclusions drawn from the results of the socioeconomic assessment and ANTECO's experience in operating the project and in implementing the EC-PSP model.

Assessing the impact of energy projects, particularly at the end-user level is important for ADB. By sharing knowledge of project impacts, ADB can support its developing member countries (DMCs) in decision-making and policy development. The knowledge about success factors of innovative projects can foster implementation of best practices by DMCs and serve as benchmark for further assistance. For NEA and the electric cooperatives, the assessment and lessons learned from ANTECO's experience would likewise help in determining the replicability of this intervention in other parts of the country. There are 121 electric cooperatives in the country and many of these have remote islands and isolated barangays within their franchise areas. The application of distributed renewable energy systems (DRES), such as the solar PV hybrid system would be valuable as a key component to address the energy access, energy security, and reliability challenges of these electric cooperatives.

The output of the study will enhance knowledge about DRES and its impact on rural communities. The lessons learned from the project may be used to further improve project implementation, encourage replication, and increase DRES deployment within the country and beyond.

2 THE MALALISON ISLAND SOLAR PHOTOVOLTAIC HYBRID PROJECT

Project Rationale and Objectives

Globally, there has been an increasing trend in the use of DRES to generate electricity, either as a stand-alone system or as a part of a microgrid or mini-grid. Populations that are not served through grid connections or extensions can be provided access to electricity through DRES solutions using renewable energy resources such as solar PV, bio-power, wind, and hydropower. This strategy is seen as a solution to facilitate economic development, help curb the worsening climate change, and improve climate resilience.

The diesel generation facilities deployed in small islands in the Philippines may be retrofitted or hybridized with renewable energy technologies based on availability of these resources in each area. This option would enable existing mini-grids to supply sufficient and reliable 24-hour electricity and at the same time reduce subsidies as the use of expensive fossil fuel is reduced. For unelectrified areas, the deployment of mini-grid installations using renewable energy hybrids also appears to be a sound and logical idea from a long-term development perspective. Considering that 91% of off-grid power supply relies on fossil fuel, the use of renewable energy hybrid mini-grids would surely impact positively on national energy security, enhance climate resilience, and reduce poverty incidence.

A recent analytical study released by the Institute of Energy Economics and Financial Analysis (IEEFA) on the Philippines off-grid subsector supports these aforementioned conclusions. According to the study, renewables are now more affordable, reliable, and resilient as the solution for small island and isolated power grids. Based on IEEFA's analysis, "solar PV plus lithium-ion batteries can now reliably deliver power at a significant discount to the price-performance potential of the current diesel-power fleet." The analysis shows that in an optimized scenario, renewable energy even at a low of 30% share can deliver power at 18% lower than a diesel only option. The study further notes that "considering the deflationary trend of renewable energy and the volatility of oil prices to international market prices, renewable energy offers small island and isolated grids both low prices and price stability." A shift by NPC-SPUG away from diesel would realize savings up to ₱13.5 billion ($275 million) for the government in terms of reduced subsidies.[4]

NEA, the government agency in-charge of the country's rural electrification program, supervises and provides financing and grant subsidies to the 121 electric cooperatives to deliver electricity services in rural areas, including small and isolated islands within the electric cooperatives' franchise coverage. Among these are 39 electric cooperatives in small and isolated islands. NEA wants to undertake a more programmatic approach to provide sufficient, reliable, and clean electricity service to these islands.

[4] S. J. Ahmed. 2020. *Renewables Are a More Affordable, Reliable, Resilient Solution for Small Island and Isolated Power Grids.* Institute for Energy Economics and Financial Analysis.

In 2015, ADB introduced the idea of hybridizing power generation within the franchise areas of electric cooperatives. ADB, with the support of Korea Energy Agency, conducted five feasibility studies and eventually pilot-tested the solar PV hybrid technology in Cobrador Island, Romblon Province, in collaboration with NEA and Romblon Electric Cooperative.[5] This strategy of deploying solar hybrid systems was deemed applicable for replication in other islands in the Philippines. It was contemplated that aside from replicating the technology, a new pilot project would take a step onward to commerciality by testing the idea of bringing in private sector investment using EC-PSP model and introducing the prepayment or prepaid metering technology in an off-grid setting.

Through the EC-PSP model, NEA hopes to encourage investments by the private sector in off-grid electrification. This EC-PSP model is essential to cover the government's funding gap in its quest to attain total electrification, to help reduce subsidy as well as lower end-user tariff. This strategy is consistent with the policy framework set by the government of adopting privatization as the country's strategy in the power sector.

Specific to Malalison Island, the pilot project has three main objectives:[6]

(i) Capacity expansion to extend service hours from 4 to 24 hours per day and improve the quality of service;

(ii) Pilot-testing of prepaid metering system in an island setting to improve collection efficiency and reduce billing and collection costs; and

(iii) Pilot-testing of EC-PSP business model.

Selection of Project Site

Selection Criteria

During project conceptualization, ADB contemplated on developing a pilot project in a location affected by Typhoon Haiyan, but outside of the Samar–Leyte area where help was most concentrated. In consultation with NEA, Panay Island was targeted and the following off-grid sites were considered: Barangay Malalison, Culasi, Antique; Barangay Botlog, Concepcion, Iloilo; Barangay Bayas, Estancia, Iloilo; Sitio Naburot, Barangay Buenavista, Carles, Iloilo; Barangay Manlot, Carles, Iloilo; and Barangay Nasidman, Ajuy, Iloilo.

To select the pilot project site, ocular inspections of the potential areas were conducted, and each potential site was subjected to a set of criteria as follows:

- **Need and potential for mini-grid expansion in the area.** The pilot project should be in an area that is either not served or underserved, and has sufficient demand and market base that would require a mini-grid.
- **Number and configuration of households.** The site should have around 200 households. Areas that are too small might have little impact while sites that are too big will require more investment and are not ideal for a pilot project.

[5] ADB. 2019. *Guidebook for Deploying Distributed Renewable Energy Systems: A Case Study of the Cobrador Hybrid Solar PV Mini-Grid.*

[6] Letter of Agreement between ADB, NEA, and ANTECO. September 2016 (unpublished).

- **Sufficiency of renewable energy resource.** While the default is to use solar energy, other renewable energy resources, such as hydro, wind, or biomass will also be considered.
- **Accessible location.** The location for pilot should be accessible enough to facilitate construction and/or installation and to demonstrate results and outcomes to other areas or regions.
- **Acceptability** by stakeholders, such as, local government unit (LGU) and local leaders.
- **Proven technology solutions.** The pilot project is not a technology development program and should not support activities or projects that involve unproven technologies or integration.
- **Electric cooperative capabilities.** The electric cooperative should have demonstrated capacity and resources to make projects happen quickly. They should be carefully vetted to avoid unmanageable delays. An electric cooperative that has the capability to finance the project is preferred. ADB will provide limited grant funding and should leverage financing from the electric cooperative and the private sector.

Based on the selection criteria, Barangay Malalison, Culasi, Antique Province was selected as the pilot site.

The Project Site

Barangay Malalison (also known locally as Mararison) is a fishing village located in Malalison Island, municipality of Culasi, Antique Province. It can be accessed in about 15–20 minutes by motorized boat off the coast of Culasi. After Typhoon Haiyan, the barangay had 153 households. Several households temporarily relocated to the mainland while rebuilding their houses that were destroyed by the onslaught of the typhoon. The people's main occupation include fishing, boat making, and local tourism. The barangay has basic facilities including a barangay hall, an elementary school, health station, and a daycare center.

Malalison Island. Malalison is a hook-shaped island off the coast of Culasi municipality, in the Province of Antique, Philippines (photo from Municipality of Culasi, Antique).

Table 1 presents the general information on Malalison Island in 2015 during the inspection visit made by ADB consultants.

Table 1: Malalison Island—General Information (2015)

Area Description	An island barangay
Land Area	55 hectares
Number of Households	153 (after Typhoon Haiyan)
Means of Access	About 20 minutes by motorboat from town proper of Culasi
Livelihood	Fishing, boat making, tourism (food stalls and small lodging)
Barangay Facilities and Services	Barangay hall, elementary school, health station, daycare center; huts for rent, spring water collected and delivered to residents
Energy Profile	Energized in 2013 using a 25-kilowatt genset operated by ANTECO: • Hours of operation—4 hours (6–10 p.m.) • Distribution line—single-phase; 13.2 kilovolts • 148 households availing electricity service from ANTECO diesel genset • Tariff is ₱25/kilowatt-hour (kWh) ($0.53)[a] subsidized by electric cooperatives • True cost is ₱35/kWh ($0.74)[a] • Some households use solar lanterns

ANTECO = Antique Electric Cooperative.

[a] $1 = ₱47.

Source: Annex B: Malalison Solar PV Hybrid Project Signed Letter of Agreement.

The island barangay is within the electricity distribution franchise of ANTECO. It is underserved in terms of electricity supply. Power is supplied by ANTECO using a 25-kilowatt (kW) diesel generator that provides the island limited 4 hours of electricity, from 6–10 p.m., daily. Due to the fluctuating and relatively high cost of diesel fuel and the burden of transporting fuel to the island, people in the island have to pay a much higher tariff for their limited service than those in the mainland, where electricity is available 24 hours daily. To make the tariff more affordable, ANTECO partially subsidies the cost of power generation.

In terms of its acceptability as the pilot site, it was noted that Barangay Malalison complied with the requirements of ADB's selection criteria:

- The barangay is underserved with only 4 hours of electricity service daily.
- There appears to be a market base to expand the existing mini-grid service of ANTECO. Based on interviews with local officials, the number of households is expected to increase to about 200 and there is a good potential for demand to increase from the island's local tourism industry. Malalison Island with its white beaches is considered as an alternative to Boracay among local tourists and its terrain is similar to Batanes.
- The island is accessible. It is only 15–20 minutes by boat from the mainland. Logistically, it would not be too difficult to transport materials to the island during construction, though the challenge of a beach landing persists, which is common in small islands. Moreover, as a showcase of an innovative technology, the island is accessible for study visits and knowledge sharing with other electric cooperatives and interested parties.

- The local officials, both at the barangay and municipal levels, have shown keen interest and manifested their readiness to support the project.
- ANTECO and its management were willing to undertake the project. This willingness was backed up with a board resolution signed by all its board members formalizing the electric cooperative's commitment to implement the pilot project and provide the necessary counterpart financing.

Based on these findings and with the endorsement of NEA, ADB approved the project. In November 2016, a letter of agreement was signed between ADB, NEA, and ANTECO for the implementation of the pilot project. In accordance with NEA rules, ANTECO conducted a competitive selection for a private sector partner, which resulted in the selection of One Renewable Energy Enterprises, Inc. (OREEi), a company with a track record in renewable energy project implementation.

Technology Choice and Management Solution

Considering its mandate to provide total electricity access, ANTECO faced the major challenge of finding an innovative technology and management solution to provide sufficient and reliable 24/7 electricity service to the island and improve its operational efficiencies. As experienced in other islands in the Philippines, the provision of a 24-hour electricity service is expected to open the island's development opportunities, ease the burden of women, improve children's education, among others, and by this, achieve the government's goals of poverty reduction and inclusive growth. After conducting the relevant studies, the adoption of a hybrid solar PV facility with energy storage system and back-up diesel generator was deemed the appropriate solution. ANTECO's operations will be enhanced to deliver improved services with the use of a prepayment metering system.

Aerial view of the Pilot Malalison Solar Photovoltaic Hybrid Power Plant. The project was installed to provide the island with sufficient and more reliable power supply (photo by Asian Development Bank).

Resource Assessment

Solar power was selected for the pilot project because it is the only resource found to be abundant and feasible in the island. Upon conducting a survey of existing renewable energy resources, the island was found to have sparse vegetation with grassy hills. Arable land is limited, with coconut trees scattered around the island. The wind regime was not enough for power generation and no hydro resource was found. On the other hand, global solar horizontal irradiation data from the United States National Aeronautics and Space Administration and National Renewable Energy Laboratory was determined to be at 5.07 and 5.00 kilowatt-hour (kWh)/square meter (m^2)/day annual average, respectively, with monthly averages of over 4 kWh/m^2/day (Table 2).

Table 2: Global Irradiation Data in Malalison Island

Month	Global Horizontal Irradiance (kWh/m^2/day)	
	Source: NASA	Source: NREL
January	4.71	4.38
February	5.40	5.11
March	6.11	5.68
April	6.49	5.74
May	5.73	5.27
June	4.95	4.90
July	4.59	4.95
August	4.55	5.02
September	4.81	5.12
October	4.72	4.81
November	4.48	4.45
December	4.28	4.51
Average	**5.07**	**5.00**

kWh = kilowatt-hour, m^2 = square meter, NASA = National Aeronautics and Space Administration, NREL = National Renewable Energy Laboratory.

Sources: One Renewable Energy Enterprises, Inc. 2015. Technical and Financial Study. Unpublished; and, the United States National Aeronautics and Space Administration and National Renewable Energy Laboratory.

The irradiation data shows that there is sufficient solar resource to support the operation of a solar PV hybrid plant in Malalison Island.

Demand Assessment and Forecasting

In 2015, during the conduct of feasibility study by ANTECO and its private sector investor, OREEi, ANTECO was operating at only 4 hours daily, which is sometimes extended to 6 hours during special events (barangay fiesta, celebrations, etc.) and seasons (Christmas or summer). Based on the sum of the kWh readings of all individual household meters (117 households), total demand for the year was computed at 13,009 kWh or an average monthly demand for electricity of 1,084 kWh. This accounts for 9.26 kWh per month or 0.31 kWh per day, per household connected to ANTECO's mini-grid at the time of the study. For the same year, the diesel generator registered 22,189 kWh on its mother meter. This represents a monthly average of 1,849 kWh of energy generated or an average of 15.80 kWh per household per month. This indicated that monthly household kWh consumption is on the average only about 58.6% of total power generated (Figure 2). The remaining energy generated were nonrevenue and considered part of the plant's system use and distribution losses.

Figure 2: Malalison Power Supply-Demand Situationer—Household Demand vs. Energy Generated (in kWh 2015)

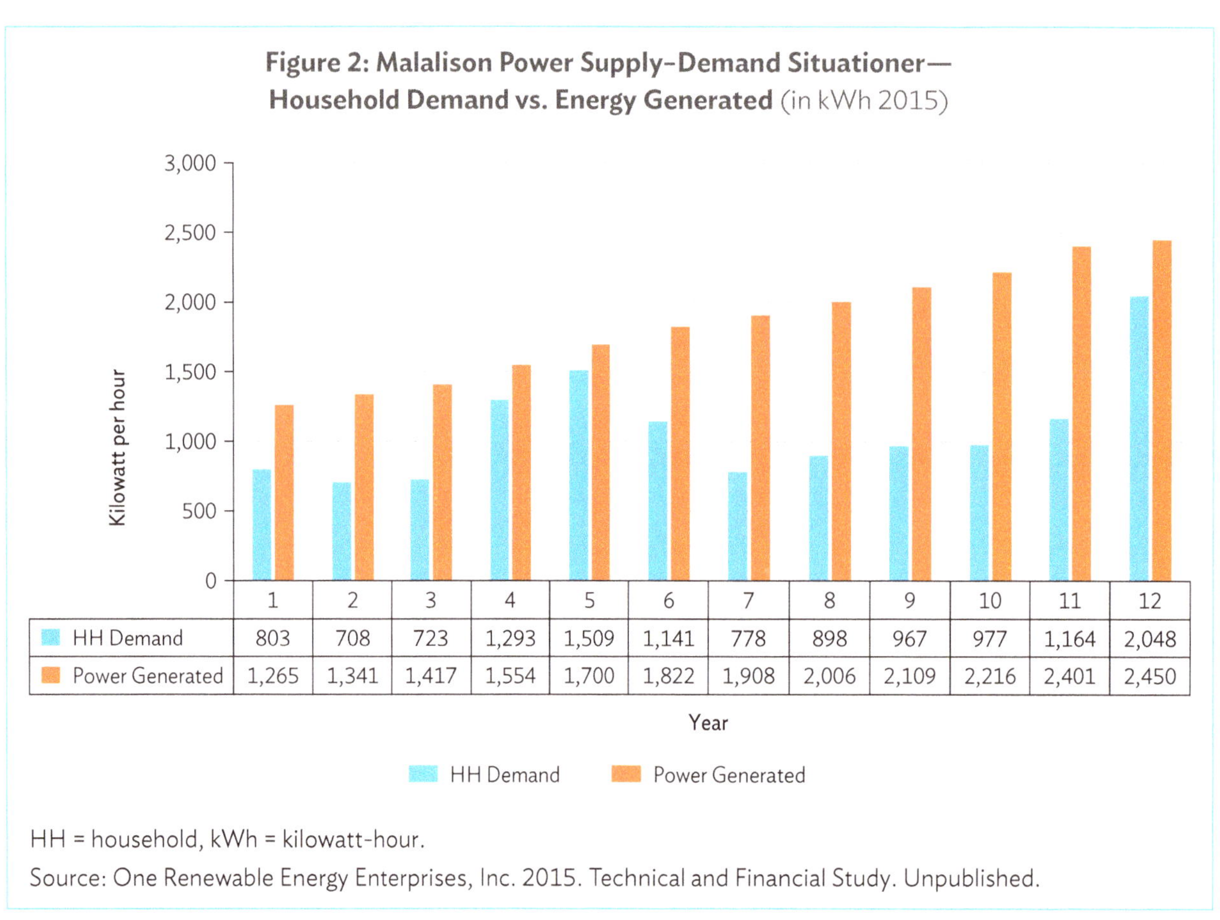

	1	2	3	4	5	6	7	8	9	10	11	12
HH Demand	803	708	723	1,293	1,509	1,141	778	898	967	977	1,164	2,048
Power Generated	1,265	1,341	1,417	1,554	1,700	1,822	1,908	2,006	2,109	2,216	2,401	2,450

HH = household, kWh = kilowatt-hour.

Source: One Renewable Energy Enterprises, Inc. 2015. Technical and Financial Study. Unpublished.

On the other hand, based on an actual survey conducted in the island, the load requirement of a typical household in the island was computed at 0.38 kWh per day or 10.70 kWh monthly, which is about 15.8% higher than the actual demand recorded for the year. Table 3 shows the estimated demand requirement per household based on OREEi's household survey considering a maximum of 6 hours of use per day.

Table 3: Estimated Household Energy Requirement

Load Items	Wattage	Number of Units per HH	Hours of Use per Day	kWh Consumption per Day
Light bulb	10	2	6	0.12
Appliance: Radio/TV/Fan	40	1	6	0.24
Charging of battery-operated devices (cellular phone, flashlight)	3	1	6	0.02
			kWh HH per day	**0.38**
			kWh per HH month	**10.70**

HH = household, kWh = kilowatt-hour.

Source: One Renewable Energy Enterprises, Inc. 2015. Technical and Financial Study. Unpublished.

Since the solar PV hybrid mini-grid system will be delivering 24 hours of electricity daily, the current average daily energy demand was assumed to increase by around 2.5 times due to suppressed demand. Increase in demand assumption also considers use of new appliances that people desired to purchase as soon as 24-hour electricity becomes available based on the household survey conducted. Thus, energy demand for year one of the project was projected to be around 243.60 kWh per day or 7,308 kWh monthly for a total of 88,914 kWh for year one, with allowance of approximately 20% system and distribution losses. Table 4 presents the assumed energy demand for year one of the project. Onward, the projected energy demand of the island was assumed to escalate at the following rates:

Year 2 and Year 3 : 10%

Year 4 and onward : 3%

Table 4: Projected Energy Demand, Year 1

Projected Number of Households/Consumers		**200**
Average Energy Demand per Household	Daily	1 kWh
	Monthly	30 kWh
	Yearly	365 kWh
Average Total Energy Demand (including System and Distribution Loss)	Daily	243.60 kWh
	Monthly	7,308 kWh
	Year 1	**88,914 kWh**

kWh = kilowatt-hour.

Source: One Renewable Energy Enterprises, Inc. 2015. Technical and Financial Study. Unpublished.

Accordingly, the projected demand for the project will increase from 88,914 in year 1 to 126,613 by year 10. Figure 3 presents the 10-year demand forecast assumed for the project.

Figure 3: Malalison Solar Photovoltaic Hybrid Mini-Grid Project 10-Year Load Forecast, kWh

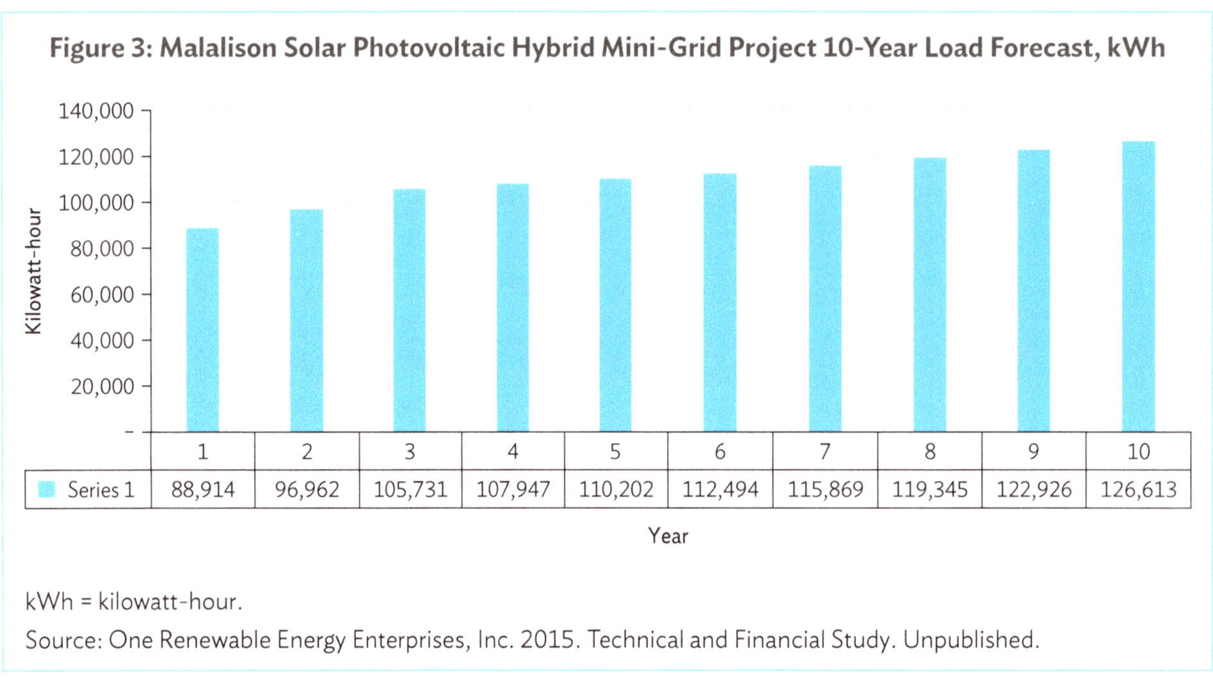

	1	2	3	4	5	6	7	8	9	10
Series 1	88,914	96,962	105,731	107,947	110,202	112,494	115,869	119,345	122,926	126,613

kWh = kilowatt-hour.

Source: One Renewable Energy Enterprises, Inc. 2015. Technical and Financial Study. Unpublished.

Since ANTECO did not have hourly kW demand data available, a load profile was assumed for the power plant referencing the load pattern of Cobrador island and adjusted to consider higher loads during the tourist season months of April, May, and December. These months were observed to be peak seasons in the island, hence there would be considerably higher electricity demand at this time, while the rainy months of July to October were considered as lean months.

Figure 4 shows the expected load profile for the project.

Figure 4: Malalison Assumed Load Profile

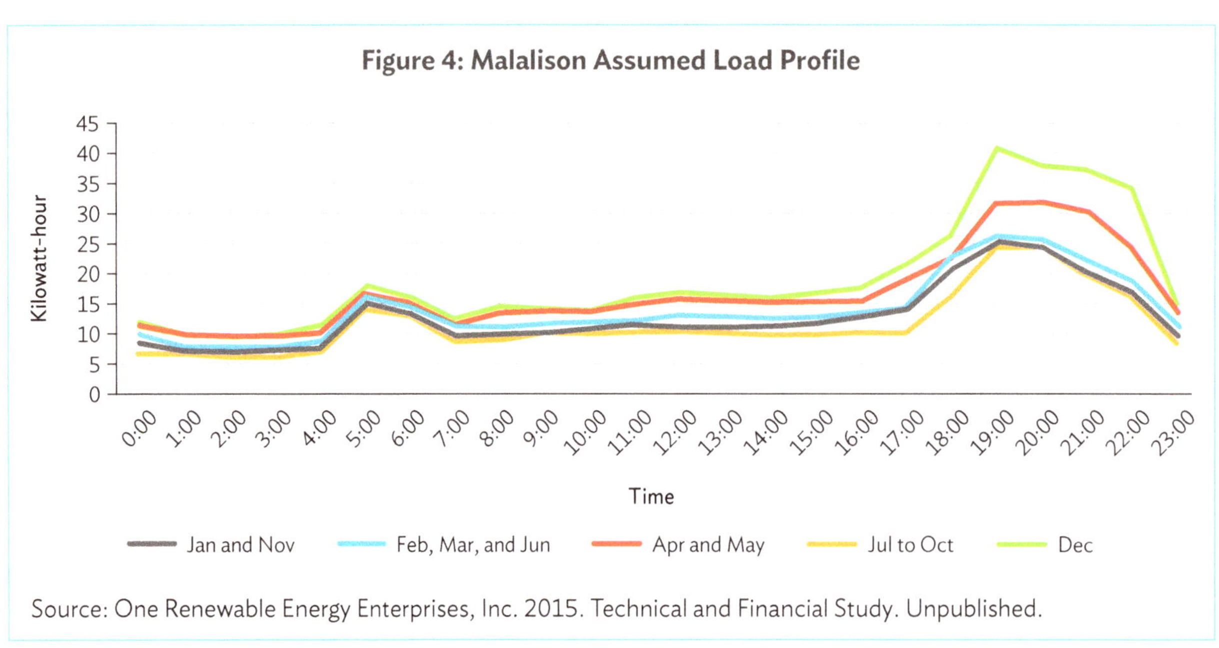

Source: One Renewable Energy Enterprises, Inc. 2015. Technical and Financial Study. Unpublished.

Technical Assessment and Design

Design Optimization

The technical design for the project was developed using the Hybrid Optimization of Multiple Energy Resources (HOMER) software. Based on the optimization runs, the solar hybrid mini-grid power plant was configured to consist of a 50-kW PV system with 273-kWh lithium-ion batteries and 54-kW diesel generator designed to produce 200 kW 24/7 power. The capacity was calculated to supply the current and future needs of some 200 households and commercial establishments catering to local tourism within the next 5 years, after which, expansion of the generation capacity should be expected.

Results of the simulations conducted by OREEi showed that the solar PV hybrid system, at the capacity earlier described, is more cost-effective, when compared with running a diesel generator-only system, as shown in Table 5. The initial capital cost for a diesel generation option is much lower than solar PV hybrid. However, the operating costs are more than double due mainly to the cost of fuel. Over the 25-year HOMER simulation, the levelized cost of energy of the solar hybrid system is lower at ₱26.33 ($0.56) compared to ₱37.99 ($0.81) if diesel is used to generate power to service the island's needs, 24/7 daily. Moreover, carbon dioxide (CO_2) emissions of the diesel generation option is over eight times higher than if the solar PV hybrid system is used. The solar PV hybrid option was therefore adopted as the technology of choice for the island.

Table 5: Hybrid Optimization of Multiple Energy Resources Simulation—Diesel Generator vs. Solar Hybrid System

Cost Summary (Net Present)	Diesel Generator Only (2 X 54 kW)		Hybrid System (Solar PV, BESS, Back-up Genset)	
	₱	$	₱	$
Capital Cost	2,220,000.00	47,234.00	14,339,316.00	305,091.83
Fuel Cost	22,731,151.86	483,711.81	2,728,504.00	59,053.28
O&M	8,202,901.24	174,529.81	8,385,133.94	178,407.11
Total System Cost	41,043,823.86	873,272.85	28,444,660.00	605,205.53
Levelized COE (₱/kWh)	37.99	0.81	26.33	0.56
Fuel (liters/year)	46,088		5,531	
CO_2 (kg/year)	120,651		14,840	

Exchange rate: $1 = ₱47.

BESS = battery energy storage system, CO_2 = carbon dioxide, COE = cost of energy, kg = kilogram, kW = kilowatt, kWh = kilowatt-hour, O&M = operation and maintenance, PV = photovoltaic.

Source: One Renewable Energy Enterprises, Inc. 2015. Technical and Financial Study. Unpublished.

System Architecture

As designed, power is to be principally generated by the solar PV system, stored in the batteries and delivered to the households using the transmission and distribution lines of ANTECO. A diesel generator will be on standby as back-up power whenever the power stored in the batteries become insufficient.

This will significantly reduce the use of the diesel generator as it will be used only as the secondary source of power in case of insufficient solar generation. This technology, using solar PV with battery storage, backed up by a diesel generator will assure supply sufficiency and reliable 24-hour service in the island.

The process of power generation and distribution by the solar PV hybrid system is shown in Figure 5 and the details of the project's system architecture is graphically shown in Figure 6.

Figure 5: Malalison Power Generation and Distribution System Process Flow

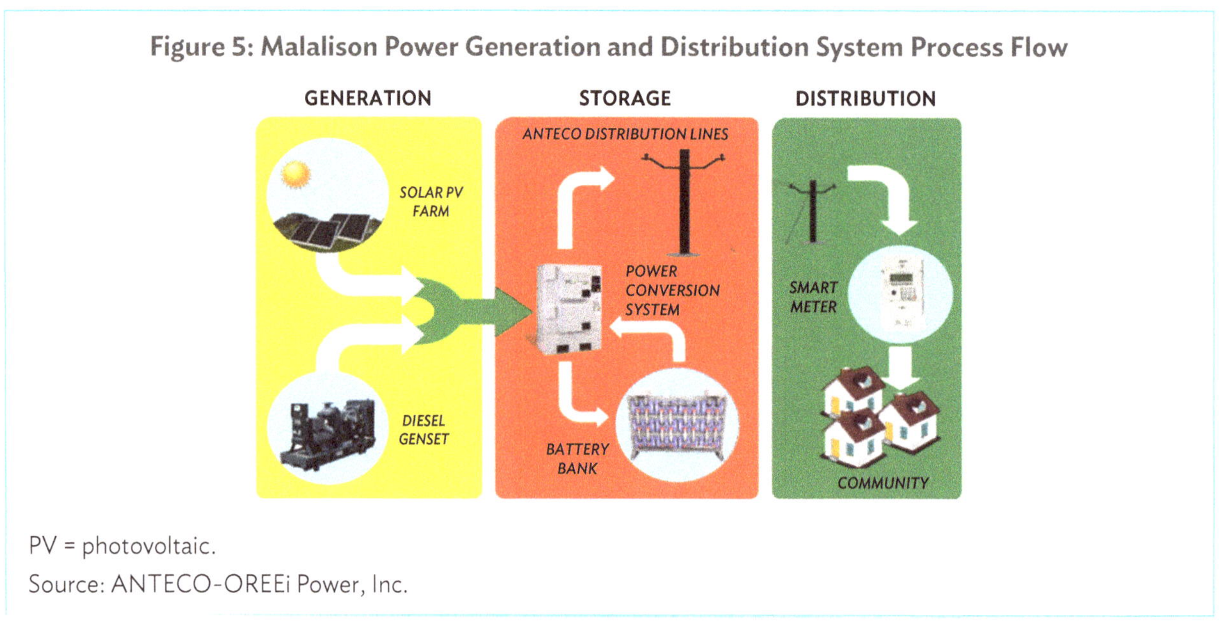

PV = photovoltaic.

Source: ANTECO-OREEi Power, Inc.

Figure 6: Solar Photovoltaic Hybrid System Architecture

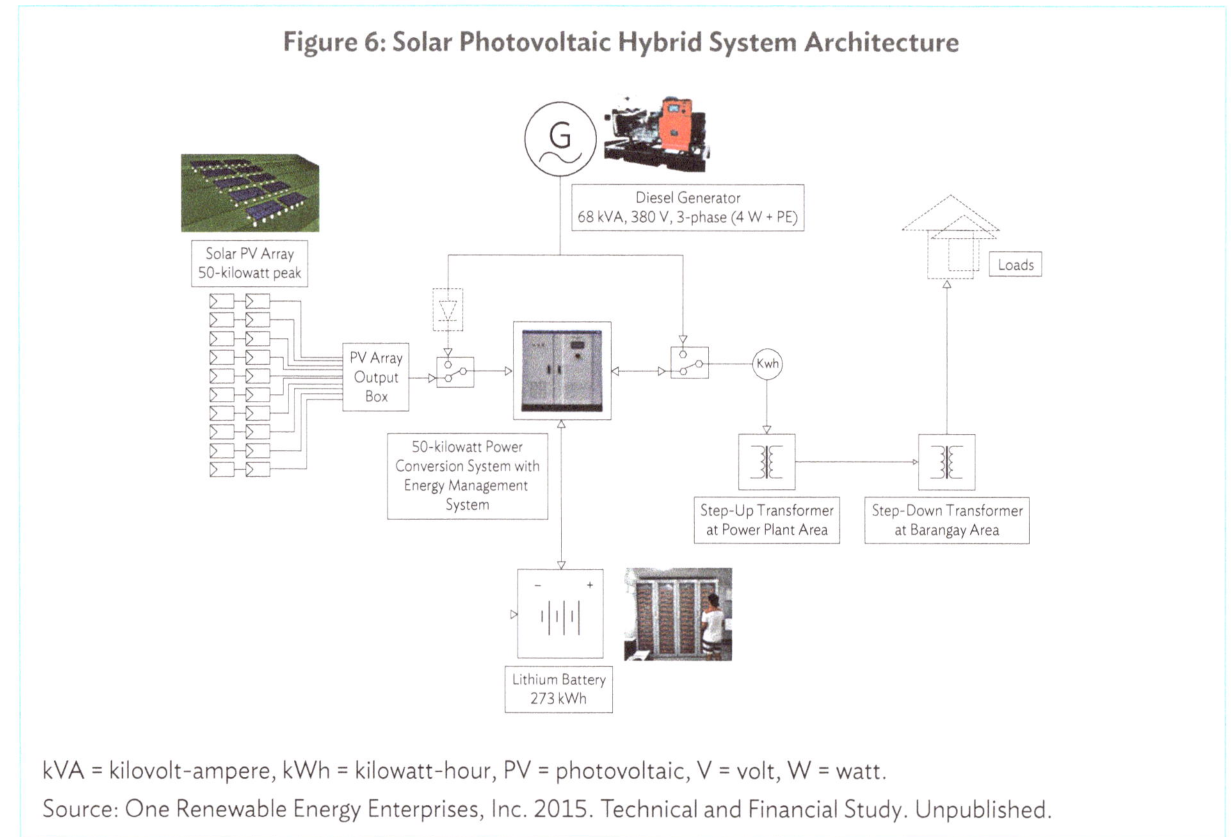

kVA = kilovolt-ampere, kWh = kilowatt-hour, PV = photovoltaic, V = volt, W = watt.

Source: One Renewable Energy Enterprises, Inc. 2015. Technical and Financial Study. Unpublished.

Solar Photovoltaic System Components

The major components of the solar PV hybrid system include:

Solar Photovoltaic Module

Several PV modules will be stringed together to form an array. For this project, multi-crystalline-type modules were used. Since the amount of electricity generated by a PV module is directly proportional to the intensity of sunlight striking the PV module's surface, PV modules should be mounted on areas where there would be no obstructions that may cause shading on the PV modules' surface. In the Philippines, the best mounting position for a PV module is to have it installed facing south with a slope of around 10–15 degrees. This is the optimal position in which the PV modules could generate the most electricity for a whole year.

Solar Mounting Structure

For this project, the solar PV modules are ground-mounted. The mounting structures are made up of aluminum frames and hot-dip galvanized steel posts. These were assembled on site and secured on concrete foundations. The project site is in Zone II of the island—with basic wind speed of 200 kilometers per hour (kph). Considering that the area is within the country's typhoon path, the mounting structures for the project were designed to withstand up to 250 kph wind speed.

Power Conversion System with Energy Management System

The overall operation of the solar hybrid system is to be controlled and managed by the power conversion system (PCS) and energy management system (EMS), which takes care of the following operations, among others:

- Charging of the energy storage system (ESS) by the PV array
- Conversion of DC electricity from the PV array to AC electricity; distribution and supply of the power to the connected AC loads through ANTECO lines
- Conversion of DC electricity from the ESS to AC electricity and its supply to the connected AC loads
- Prevention of overloading and overheating of the system
- Protection of PV array from reverse current
- Maintenance of correct and safe voltage and frequency ranges of AC output going to connected loads.

The EMS also enables the monitoring of real-time data and controls the overall operations of the hybrid system. It allows automatic switching from the ESS to diesel as and when necessary.

Energy Storage System – Lithium-Ion Battery

Consistent with ADB's energy efficiency and environmental sustainability objectives, the project uses lithium-ion batteries. Compared with conventional lead acid-based batteries, lithium-ion batteries have the following advantages:

- Higher energy density – a 1 kWh lithium battery is smaller and lighter than a 1 kWh lead acid battery
- More efficient charging and discharging – less energy losses during charging and discharging compared to lead acid

- Longer life cycle – this means longer life span of lithium batteries against lead acid
- Cleaner technology – much safer to the environment as against lead acid batteries
- Lower true cost of ownership – although initial cost of lithium batteries are much higher, they have a much longer economic life

The ESS stores excess electricity from the solar PV array during daytime and discharges this stored energy at night for use by the connected loads. It also stabilizes the voltage of the system, preventing power fluctuations that may be caused by the PV array.

The main components of the solar PV hybrid system is presented in Table 6.

Table 6: Solar Photovoltaic Hybrid System Main Components

Item	Description	Quantity
Solar Photovoltaic Module	315-watt peak, multi-crystalline	160 pieces
Solar Array Mounting Structure	Aluminum material, ground-mounted, with concrete foundation	1 lot
Power Conversion System (PCS) with Energy Management System (EMS)	PCS: 50 kW, Bi-directional Inverter, DC/DC converter[a] EMS: For monitoring and control of the overall power plant system	1 unit
Energy Storage System	Lithium battery, 273 kilowatt-hour, with built-in battery management system	1 lot
Diesel Generator	68 kilovolt-amperes, 380-volt, 3-phase (4W+PE), with digital control system	1 unit
Powerhouse/Control and Battery Room		1 lot
Other Electrical Equipment, Devices, and Accessories	Automatic transfer switch, charger, uninterruptible power supply, cables, conduits, grounding materials, etc.	1 lot
Transmission and Distribution, Equipment, Devices, and Accessories	Transformers, cables, energy meter, etc.	1 lot

[a] DC/DC converter converts DC output from lower voltage (batteries) to higher voltage (power transmission).

Source: One Renewable Energy Enterprises, Inc. 2015. Technical and Financial Study. Unpublished.

Control and Battery Room and Powerhouse

The Malalison Solar PV Hybrid Project was constructed on top of a hill located at Zone II of the island in a piece of government property donated by the local government of Culasi. The area is ideal for the project because it is open and unobstructed. As designed, the control and battery room is an 8,000 millimeter (mm) x 4,500 mm structure, separately constructed from the powerhouse. This is part of the safety feature adopted for the project. On the other hand, the powerhouse with the dimension of 4,000 mm x 3,500 mm stands 5,000 mm away and connected to the control and battery room through a bridgeway. The perspective of the project's Control and Battery Room and Powerhouse is shown in Figure 7.

Figure 7: Perspective of Control and Battery Room and Powerhouse

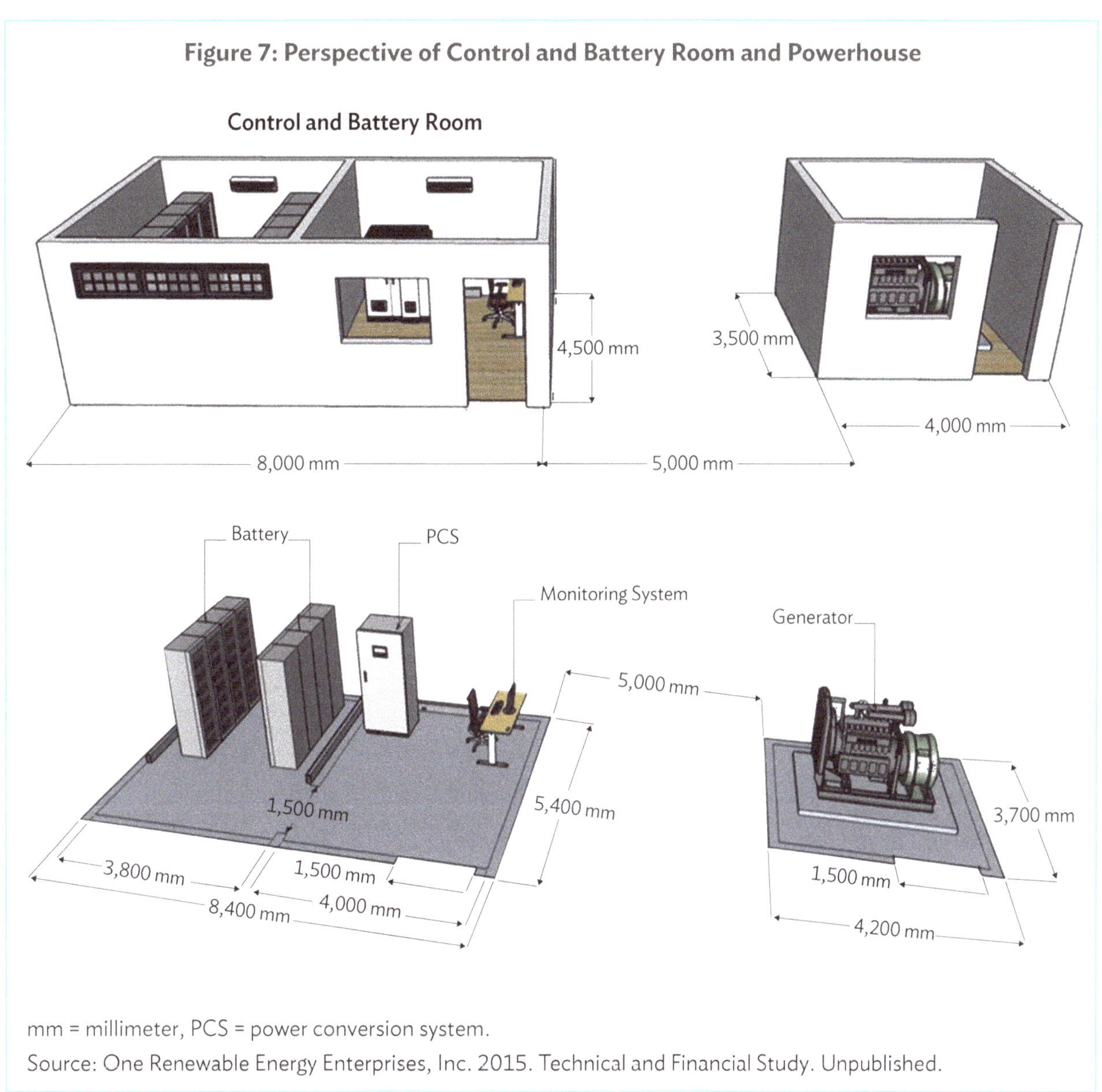

mm = millimeter, PCS = power conversion system.

Source: One Renewable Energy Enterprises, Inc. 2015. Technical and Financial Study. Unpublished.

Prepaid Metering System

Aside from this innovative power generation technology, the project also pilot tested the use of a prepaid metering system to improve the efficiency of ANTECO's billing and collection system. While prepaid metering is already being utilized by large distribution companies, the use of this technology in Malalison Island was the first to be implemented in an island setting in the country. The use of prepaid metering system is considered appropriate in the Philippines because Filipinos are already used to the prepayment system offered by the telecommunication companies. The successful application of this innovative metering system by ANTECO will encourage its replication by other electric cooperatives in response to their billing and collection challenges.

The prepayment system is part of the grant provided by ADB. The system included the following components:

- Hardware (e.g., database and encryption server, credit dispensing unit workstation, encryption module, printer)
- Software (e.g., system software, prepayment software system installation, interconnect software, system service)
- 185 units of single-phase integrated keypad prepayment meter
- Two units of 3-phase integrated keypad prepayment meter

The prepayment solution is modular and adopts the standard design. It enables online prepaid electricity through the synchronous transport signal (STS) prepayment technology. This allows users to purchase electronic loads from retail outlets at their convenience.

The single-phase residential meter is used in a split prepayment metering system. It complies with STS standard and communicates with the customer interface unit by meter bus or programmable logic controller for energy consumption monitoring and credit charging. The three-phased prepayment meter, on the other hand, has migrating advanced metering infrastructure (AMI) with STS prepayment functions and can collect detailed real-time metering information.

The system can upgrade seamlessly to smart prepayment system to realize STS prepayment and AMI functions. Among the main features of the system are:

- Scheduled and on-demand meter reading; demand limiting and anti-hoarding mechanism
- Remote tariff programming
- Remote connection, disconnection, and group control
- Tamper detection, grid, and meter event detection
- Outage detection function
- Credit balance inquiry function
- Compatibility with low-speed telecommunication networks
- Concurrent mass transaction management
- High security and hack-proof design and implementation
- Zero dependence of utility to meter provider

The introduction of the prepayment metering as a component of the pilot project is designed to increase ANTECO's operational efficiency and at the same time build the private investor's confidence in the project. The following benefits may be derived from the use of the prepaid metering system:

- For the consumer, there is 100% inclusivity. All households have access to electricity service upon demand, at levels and amounts that fit their available cashflow or budget. They can monitor their consumption and plan out their energy usage. They are not subject to bill shocks, and threatened with disconnection in case of nonpayment of monthly electricity bill which happens when electricity service is postpaid.

- For ANTECO as power plant operator, there is improved operational efficiency and substantial savings. There will be no need for ANTECO to deploy people for meter reading, billing and collection, disconnection, and reconnection. Salaries and wages, travel expenses, and other related costs will be avoided.
- For the investor, there is 100% collection efficiency and therefore its return on investment is assured.

Innovative Business Model

The project showcases the feasibility of implementing a joint venture arrangement between an electric cooperative (ANTECO) and a local private investor, OREEi, which has been selected through a competitive selection process according to government rules. The entry of the private sector will provide the needed capital, management, and technical capability to accelerate the deployment of renewable energy mini-grids in other parts of the country. Private sector investment in greenfield, small, and isolated island areas will also help promote commercialization of mini-grids and eventually reduce subsidies. Currently the government provides NPC-SPUG fuel and operational subsidies through the Universal Charge for Missionary Electrification (UCME) to lower the cost of electricity and ensure affordability in off-grid areas. But the government by itself will not be able to continue funding both the infrastructure and operating requirements to achieve its total electrification goals. The participation of the private sector in off-grid electrification is therefore critical. However, bringing in private sector investments into small and isolated islands is not easy due to high commercial risks and lack of economies of scale. Thus, aside from further validating the technical feasibility of a solar PV hybrid system, this project is also expected to demonstrate how private sector can work together with an electric cooperative to fulfill the goal of economic development through clean energy deployment.

Project Ownership Structure

As envisioned, the pilot project will be initially owned by a duly registered joint venture company established by ANTECO and OREEi, known as ANTECO-OREEi Power, Inc. (AO Power, Inc.) with 55%–45% ownership stake in favor of ANTECO. The electric cooperative will continue to serve as the power distributor in the island and will be the off-taker of power generated by the solar PV hybrid power plant. The joint venture arrangement has an exit provision to allow ANTECO to eventually own the system in the future.

Role of Stakeholders

The private sector is expected to bring in technical and management capacity to enhance and assure the sustainability of the project. On the other hand, ANTECO being a nonstock, nonprofit corporation owned by the consumers, has vast experience in working with the local communities within their franchise area. Its lead role in the pilot project gives an assurance of public acceptability and people's participation in the development, implementation, and operation of the project. NEA and the local government of Culasi will provide government support to the project. At this instance, NEA provided equity financing for the project while the LGU of Culasi donated the land on which the project was installed. The LGU also facilitated the project's compliance with the government's regulatory requirements, including LGU business and environmental permits.

ANTECO managed the local participation of the people of Malalison in terms of skilled labor for construction of the project, while OREEi prepared the technical design and managed the construction of the project as well as installation of equipment. This strong participation from all stakeholders were valuable to the successful completion of the project, which was dubbed by ADB management as a truly public–private–people's partnership initiative. The people of Malalison Island provided skilled labor and logistics, including manually carrying poles and other needed materials from the beach landing to the hilltop location of the power plant during its construction.

Dagyaw sa Mararison (community in action). People's participation in bringing materials from the shore to the project site during project construction (photos by ANTECO).

Project Cost and Financing

Total Project Cost

The total cost for the pilot project amounted to ₱28,135,502 equivalent to $521,027.82, excluding the cost of the land which was donated by the local government of Culasi. This is lower than the estimated cost of $600,000, as per the ADB–NEA–ANTECO letter of agreement. The joint venture company did not access any loan for the project. Rather, ANTECO sourced a portion of its funding from NEA, while OREEi obtained a loan from the Land Bank of the Philippines, Antique Branch. Financing contribution from the major stakeholders of the project is shown on Table 7.

ADB's 38% grant contribution to the project was credited to ANTECO, bringing its contribution to 63% of total project cost, including the prepaid metering system.

Table 7: Major Contribution by Project Partners

Project Partner	Contribution	Amount[a]		Percent
		₱	$	
ANTECO	Equity	7,104,526	131,565.30	25
OREEi	Equity	10,230,976	189,462.52	37
Municipality of Culasi	Land donation: 1,500 square meters			
ADB	Grant: Energy Storage	5,400,000	100,000.00	19
Total Cost – Solar PV Hybrid Power Plant		**22,735,502**	**421,027.82**	
ADB	Grant: Prepayment metering system	5,400.000	100,000.00	19
Total Project Cost		**28,135,502**	**521,027.82**	**100**

ADB = Asian Development Bank; ANTECO = Antique Electric Cooperative; OREEi = One Renewable Energy Enterprises, Inc.; PV = photovoltaic.

Note: $1 = ₱54.

[a] Excluding cost of donated land.

Source: AO Power, Inc.

ADB Financial Assistance

After due diligence, ADB, upon the request of NEA, agreed to support the project with a partial grant not to exceed $200,000 equivalent to 40% of total project cost.[7] The grant covered:

- $100,000 for ESS, as partial subsidy, to cover the cost of 238 kWh, out of the 273 kWh, of lithium-ion batteries with battery management system, and
- $100,000 for prepayment metering system

This support from ADB was in keeping with its role in promoting social and economic progress through energy access within the Asia and Pacific region and consistent with ADB's Energy Policy 2009 (now updated 2021 Energy Policy)[8] and Strategy 2020 (now updated Strategy 2030[9]) of achieving a prosperous, inclusive, resilient, and sustainable Asia and the Pacific.

7 Letter of Agreement between ADB, NEA, and ANTECO, signed 11 November 2016.

8 ADB. 2023. *2021 Energy Policy of the Asian Development Bank: Supporting Low-Carbon Transition in Asia and the Pacific.*

9 ADB. 2018. *Strategy 2030: Achieving a Prosperous, Inclusive, Resilient, and Sustainable Asia and the Pacific.*

3 SUSTAINABLE DEVELOPMENT AND SOCIOECONOMIC ASSESSMENT FRAMEWORK

The approach to assessing the socioeconomic impact of energy projects is not straightforward. The World Bank Energy Management Assistance Program in its report, *Beyond Connections: Energy Access Redefined,* proposes moving away from the simple definition of access to energy, i.e., that of a household electricity connection, an electric pole in the village, and an electric bulb in the house. Instead, it recommends that "the distinction between access to energy supply, access to energy services, and actual use of energy must be clearly reflected in the definition of energy access."[10] This is to recognize that energy access brings about multifaceted benefits that impacts people's lives. In remote areas where most people are poor, electrification is seen to alleviate poverty, increase access to basic services, and advance social systems.[11] In an attempt to cover these perspectives, including that of climate resilience, "the definition of sustainable energy today has broadened from the primarily economic development focus in the 1970s to concerns with environmental sustainability in the 1980s, to financial sustainability in the 1990s, and finally to the current inclusion of social sustainability, equity, and poverty."[12] Moreover, with the everchanging geopolitical landscape and uncertainties in global markets, energy security has become a major concern.

Though many studies have been conducted on this subject, there is no singular framework in assessing the impact of energy access projects to sustainable development, particularly for rural off-grid island communities. In the absence of such standard framework, this study will use the World Energy Council's trilemma of energy security, energy equity, and environmental sustainability as its reference framework (Figure 8). According to the World Energy Council, a balance of the three core dimensions of energy security, energy equity, and environmental sustainability is a measure of a healthy energy system.[13]

Aside from the energy trilemma's three dimensions, the social dimension is also deemed a necessary component to be considered in the assessment framework. Social acceptance, improved social services, and livelihood enhancements are also requisite to sustainability. In the Philippine context, this is an important consideration as government begins to shift rural electrification initiatives to the private sector. Private investment necessarily expects to generate profits, which means that energy systems should enable socioeconomic growth and sustainable development of the communities to enhance demand and improve affordability. In this study therefore, the conceptual framework for impact assessment will combine the energy trilemma with the social dimension as shown in Figure 9.

---

10 Energy Sector Management Assistance Program. 2015. Beyond Connections: Energy Access Redefined. *ESMAP Technical Report.* No. 008/15. The World Bank Group.

11 L. Lozano and E. B. Taboada. 2021. The Power of Electricity: How Effective Is It in Promoting Sustainable Development in Rural Off-Grid Islands in the Philippines? https://www.mdpi.com/1996-1073/14/9/2705.

12 E. Cecelski. 2003. *Enabling Equitable Access to Rural Electrification: Current Thinking on Energy, Poverty, and Gender.* World Bank.

13 World Energy Council Trilemma Index. https://www.worldenergy.org/transition-toolkit/world-energy-trilemma-framework.

Figure 8: The Energy Trilemma

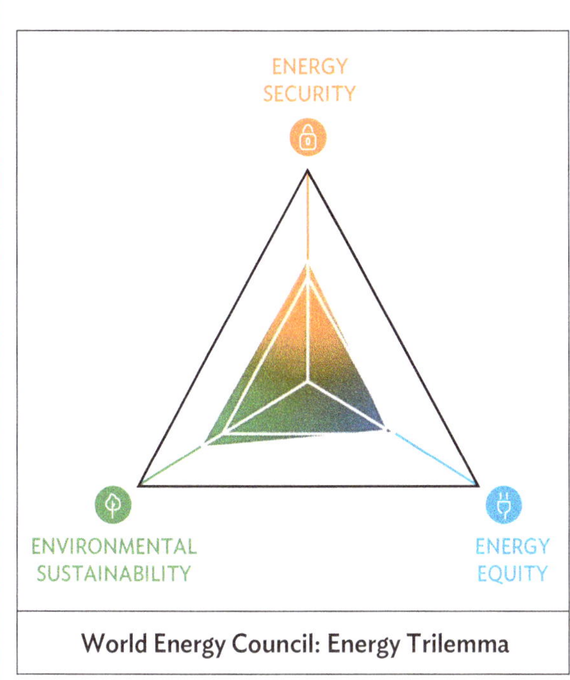

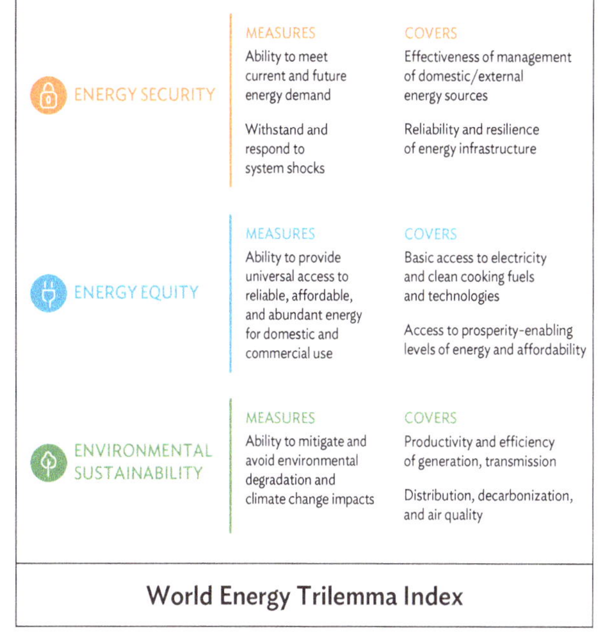

Source: World Energy Council.

Figure 9: Conceptual Framework

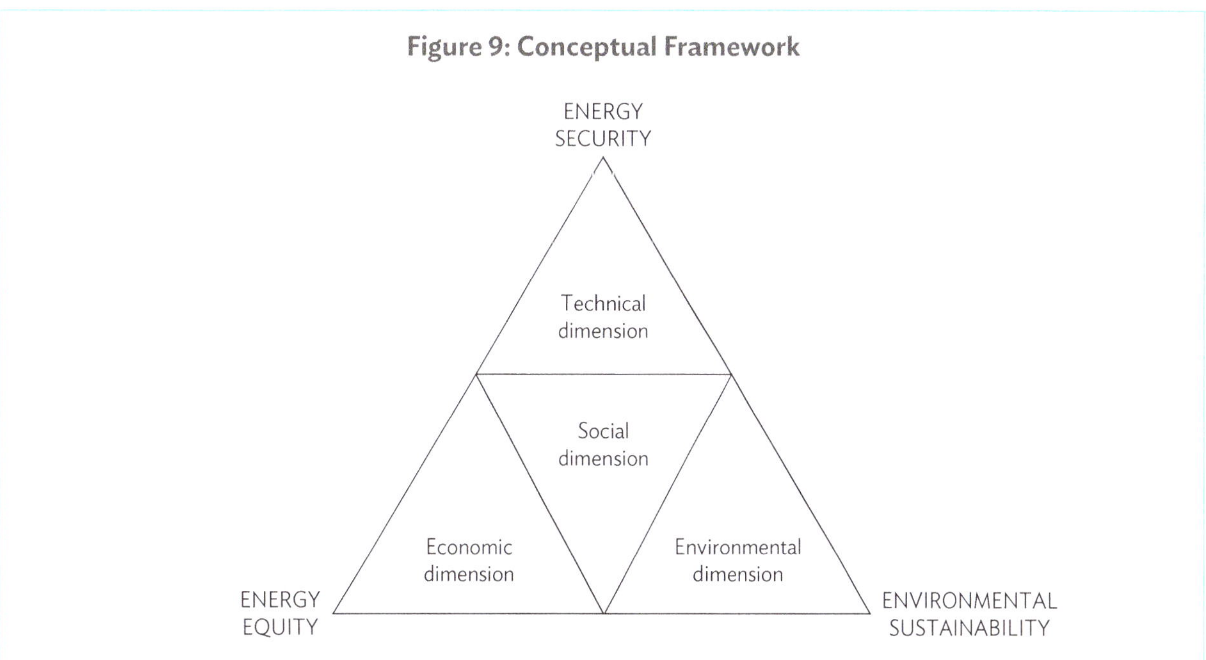

Source: Adapted from L. Lozano and E. B. Taboada. 2021. The Power of Electricity: How Effective Is It in Promoting Sustainable Development in Rural Off-Grid Islands in the Philippines? https://www.mdpi.com/1996-1073/14/9/2705.

Applying the framework to the local setting, a set of sustainable development indicators reflecting the dimensions of the energy trilemma, including social welfare factors will be used to gauge the project's impact. These indicators are presented in Table 8. The conditions at the households and the community, before and after the project's implementation, will be compared to determine changes that have occurred in the lives of the people and the community. The measurements and values representing the impact relative to each of the indicators will be generated from the results of surveys that were conducted before and after the project as well as operational data and reports from ANTECO.

Table 8: Sustainable Development Indicators

A. Sustainable Development Indicators: Households				
Indicator	**Dimension**	**Component**	**Definition**	**Measurement**
Availability of electric supply	Energy Security	Technical	Ability to connect to the grid	Number of households/ consumers connected
Adequacy of supply	Energy Security	Technical	Ability of users to use electricity for lighting and household electrical appliances	Actual electrical appliances used in the household
Duration of supply	Energy Security	Technical	Length of time electricity is available	Number of hours of service
Service reliability	Energy Security	Technical	Availability of electricity service when needed	Number and duration of power disruption experienced
Safety and security of the system	Energy Security	Technical, Economic	Electricity has not caused accidents to human or to electrical appliances	Number of accidents and/ or reports of appliance breakdowns related to electricity
Affordability of tariff	Energy Equity	Economic, Social	Ability of users to pay for electricity	Consumption level before and after the project
Support for income generation	Energy Equity	Economic, Social	Ability of users to use electricity for productive means	Number of households/ consumers using electricity for income-generating activities; reported increase in income
Personal/ household convenience	Social Welfare	Social	Ability to ease burden of work; provide other personal/household benefits	Number of hours used for household chores; number of households reporting other benefits i.e., education, communication, etc.
Displacement of conventional fuels for lighting	Environmental Sustainability	Environmental	Ability to use electricity for lighting in lieu of conventional fuels	Proportion of households who no longer use conventional fuels for lighting

continued on next page

Table 8 *continued*

B. Sustainable Development Indicators: Community				
Indicator	**Dimension**	**Component**	**Definition**	**Measurement**
Support for barangay services	Energy Equity	Social	Ability of the barangay to provide social services	Availability and quality of social services
Support for community security	Social Welfare	Social	Ability to provide a secure environment for all	Presence of street lighting and other security infrastructure
Support for income generation in the community	Energy Equity/Social Welfare	Economic	Economic improvement in the community	Number of new businesses
Support for climate resiliency	Environmental Sustainability	Environmental	Displacement from use of fossil fuel	Solar penetration rate/percent of power generation from solar facility

Source: Developed using the WEC Energy Trilemma as a framework and the study by L. Lozano and E. B. Taboada. 2021. *The Power of Electricity: How Effective Is It in Promoting Sustainable Development in Rural Off-Grid Islands in the Philippines?* as reference.

4 SOCIOECONOMIC IMPACT ASSESSMENT

Almost 3 years after the project's commissioning, a comprehensive survey of project stakeholders was conducted in Malalison Island from 8–13 October 2021 to determine the impact of the project in the community and the lives of the residents. The respondents of the survey included 149 household heads (comprising 78% of households), the barangay secretary, 3 barangay *kagawads* (council members), 12 owners of commercial establishments (4 homestay owners, 2 private resort owners, and 6 *sari-sari* [convenience] store owners), 3 members of local organizations, 2 teachers who are residents of the island, the power plant operator, and the division chief engineer from ANTECO.

In this section, the results of the survey, as well as records from ANTECO are used to determine how the project has fared in terms of providing sustainable development from the perspectives of the households and the community. From the survey responses, the project's development and socioeconomic impacts are appraised both at the household and community levels, noting significant changes in the households' and community's experiences before and after project implementation, considering the assessment framework discussed earlier. The sustainable development indicators established in Table 8 are used as a guide to establish the impact brought about by the project.

Project Setting and Demographic

Administratively, Malalison is an island barangay (village) managed by a group of elected government officials led by a barangay captain. Annually, the barangay receives a budget from the local government of Culasi for its social services to the community. Based on barangay data of 2020, the island has 827 residents, 54% of whom are male and 46% are female. There are 192 households or an increase of 25% over the 153 households in 2015. Majority of the households (89%) are single family households while 8% have two sets of family living in one household dwelling.

As in 2015, the primary source of income of household heads surveyed is fishing (59%), complemented by small businesses (23%), and other jobs include being laborers and unskilled workers, mostly in activities related to fishing and tourism, which is the growing industry in the island. Around 11% of household heads are salaried employees, i.e., barangay officials and office staff and teachers in the island's elementary school. Others work in the mainland and/or overseas (6.7%) while a few (6%) relied on their retirement pension (Table 9).

Of the 149 household heads surveyed, 30% have completed elementary school, 54% are high school graduates, and 15% were able to complete their college education. Only a very small percentage (1%) did not have formal education. Malalison has an elementary school only. Residents go to the mainland for high school and college education.

Table 9: Means of Livelihood—Head of Households

Primary Source of Income	Number of Respondents	Percentage (%)
Fishing/fish production	88	59
Small business/homestay	23	15
Employment salary	16	11
Family worker overseas or mainland	10	7
Retirement or pension	9	6
Others (laborer or unskilled worker)	3	2
Total	**149**	**100**

Source: Survey results.

Housing is an important asset to families in Malalison Island. The housing condition in the community is characterized by 47% nipa and wood, 33% in mixed wood and concrete, and 20% concrete structures. About 99% of households own the houses they reside in.

Aside from elementary education, the barangay caters to basic services through its barangay hall (office), barangay health center, and barangay day care center. A multipurpose hall is used for community activities, such as meetings, trainings, and other events. The barangay also provides water service through a community water system. This is augmented by spring water, seven dug wells for public use, and five deep wells run by private individuals. A rainwater catchment also serves as additional water source for the community. In 2022, after the project, OREEi inaugurated a water desalination project to help improve the water supply situation in the island. As ANTECO's private sector partner, OREEi was encouraged to provide additional investment in the form of the water supply as a service to the community.

Project Impact at Households

Availability and Sufficiency of Power Supply

The project's most significant benefit is in the form of availability of reliable electricity on a 24-hour daily basis compared to only 4 hours service before the project was implemented. This gave power supply sufficiency and greatly improved the quality of service received by consumers (Table 10).

Supply sufficiency is evidenced by the substantial increase in power generation the year the project became operational. Table 11 and Figure 10 show the power generated before (2017–2018) and after the project (2019–June 2023). At a 4-hour/day service, the supply of power from the diesel generator was limited to only an 20,628 kWh for 2017 and 20,868 kWh for 2018. Upon the operation of the solar PV hybrid power plant, power generation increased, averaging 90,550 kWh over 4 years from 2019 to 2022 and 49,403 kWh for the first 6 months of 2023. This huge jump in supply availability is the result of increased generation capacity from originally a 25-kW diesel genset to 50-kW PV with battery system operating 24/7, with back-up power from a 54-kW diesel genset whenever power supply from solar resource is insufficient. Power generation from the project is now sufficient for the daily needs of 200 households, which is the optimum holding capacity of the island in terms of population density.

Table 10: Power Situation Before and After the Project

Item	Before	After
Power supply source	25-kilowatt (kW) diesel generator	50 kW solar generator 278 kilowatt-hour battery energy storage 54 kW generator (back-up)
Operating hours	6:00–10:00 p.m. (4 hours)	24 hours
Peak load	10 kW	40 kW
Service capacity	Up to 145 households	Up to 200 households

Source: ANTECO.

Table 11: Power Generation, 2017–June 2023 (in kWh)

Power Source	Before Project[a]		After Project				
	2017	2018	2019	2020	2021	2022	2023[b]
Genset	20,628	20,868	19,799	17,866	30,320	16,560	9,160
Solar			70,102	81,084	57,216	69,254	40,243
Total	**20,628**	**20,868**	**89,901**	**98,950**	**87,536**	**85,814**	**49,403**

kWh = kilowatt-hour.

[a] Diesel genset operating only 4 hours per day.

[b] January to June only.

Source: ANTECO.

Figure 10: Power Generation, in Kilowatt-hours (2017–June 2023)

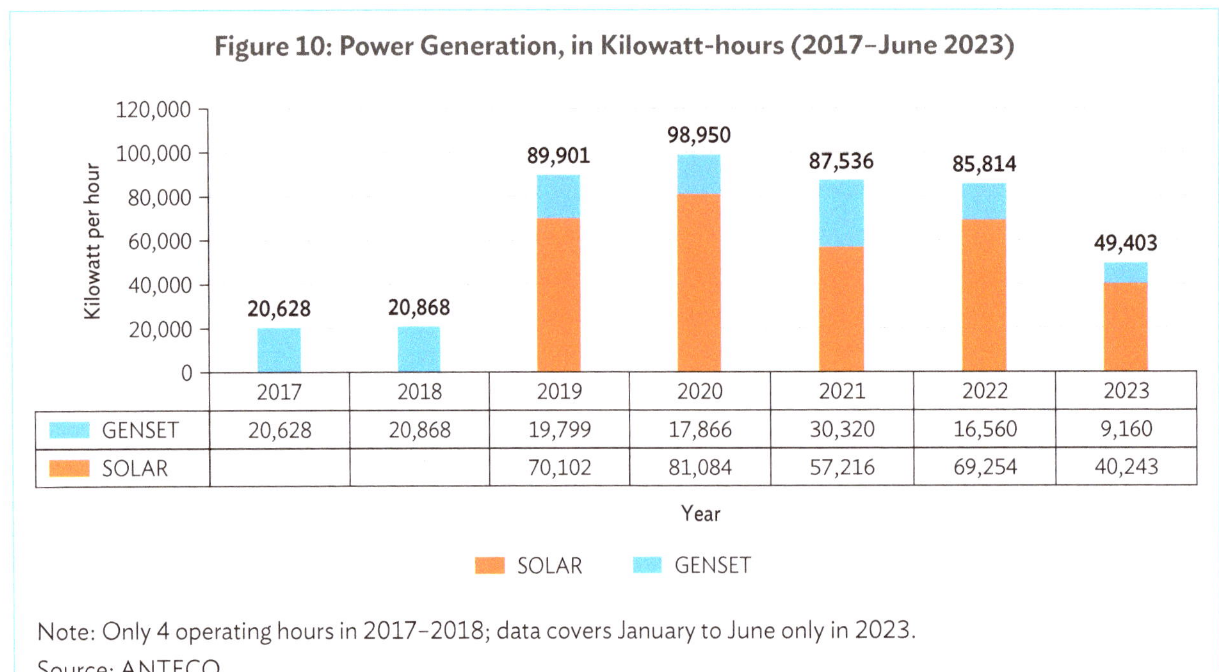

Note: Only 4 operating hours in 2017–2018; data covers January to June only in 2023.

Source: ANTECO.

Energy Access

As of 2023, ANTECO has already connected 188 consumers into the mini-grid, up from 148 in 2015 when ADB first investigated the island as potential project site. This comprises 182 or 95% of the current 192 households in the island, plus six commercial establishments. These consumers can access and receive electricity upon demand. However, not all the connected consumers actively purchase electricity all the time considering that ANTECO now operates via prepaid metering. As of June 2023, 168 active consumers were monitored as economic activities regained momentum, after the coronavirus disease (COVID-19) lockdown. Figure 11 shows the increasing trend in the number of consumers after project implementation.

Figure 11: Number of Consumers, 2017–2023

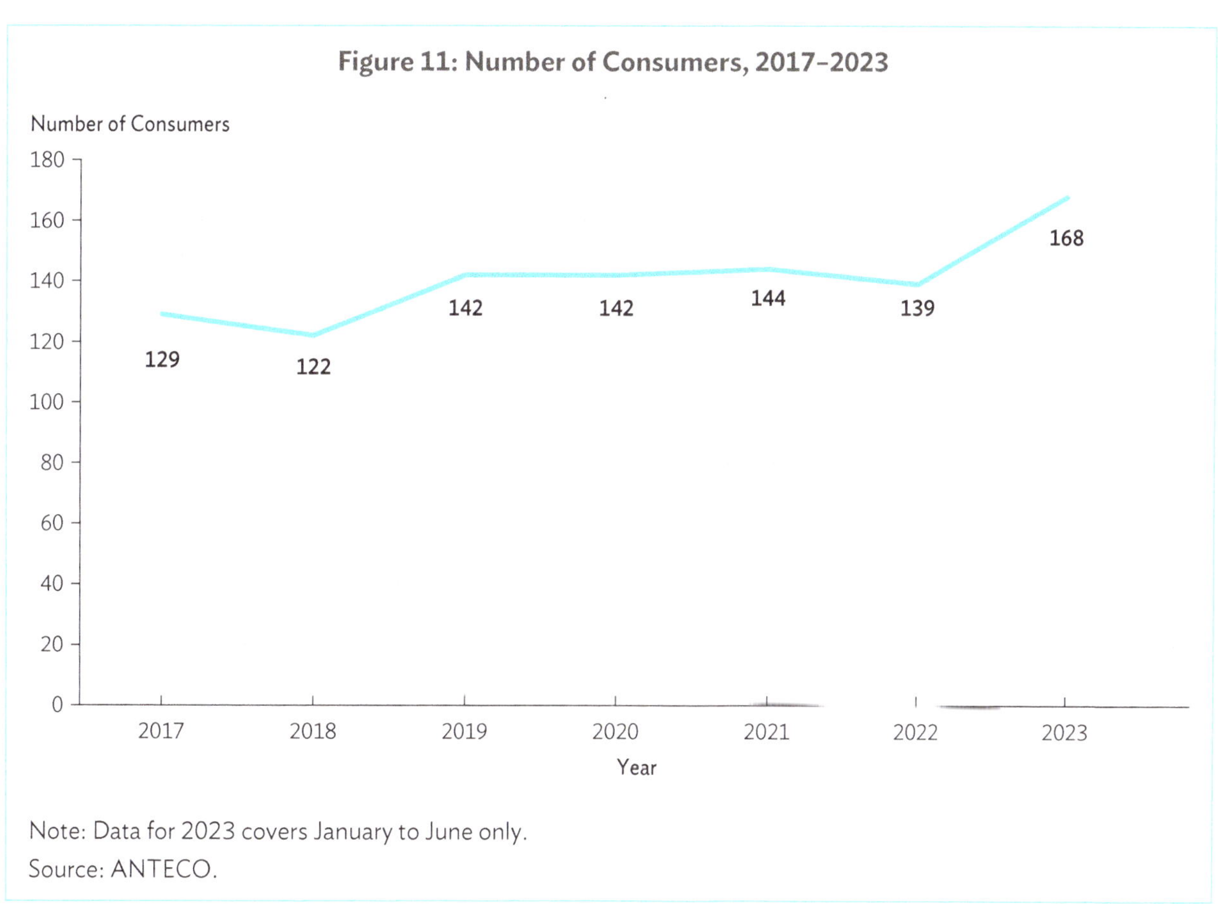

Note: Data for 2023 covers January to June only.
Source: ANTECO.

Inclusive and Efficient Energy Access

The prepaid metering system, which was introduced together with the solar PV hybrid plant, enabled more households to receive electricity service and allowed them to avail themselves of electricity whenever they need it, and at the level of consumption they can afford. Before the project, consumers had no awareness of the amount of electricity they consume. Households are billed monthly, such that, if they fail to pay, they get disconnected and need to pay the reconnection charge to get back ANTECO's service. With prepaid metering service, households can now access electricity even at minimum levels without having to worry about getting disconnected and paying for reconnection charge to get back to active status.

Previously, to be reconnected, an ANTECO lineperson had to be deployed, which will be a cost to ANTECO and causes access delay to the consumer. With the prepaid system, there is no longer any distinction between consumers who can afford regular monthly electricity billings and those who cannot. All consumers can have access to power supply any time at the amount their cashflow can afford. Even households without regular income can purchase electricity at the level and timing they prefer based on their cashflow. Households are also able to monitor their consumption. They may limit their usage by kWh and be more efficient in using electric appliances and gadgets to save on their electricity expenses. In effect, the use of prepaid metering has brought about both efficiency and inclusivity in energy access.

Household Convenience

With 24-hour electricity service, households were able to acquire and use electric appliances that they were otherwise constrained to do before the project's operation. Most of the appliances purchased are those for convenience and comfort. Notable of these are cellular phones and refrigerators, freezers, television sets, electric fans, rice cookers, and washing machines. Table 12 presents the top 10 newly acquired electric appliances by respondents in addition to what the households already have prior to the project. Respondents reportedly purchased 48 new electric fans, 47 rice cookers, and 43 washing machines. The number of cellular phones increased by 68% and television sets by 48%. Moreover, the number of refrigerators increase by 90% and 16 new freezers were also acquired while there was none before the project. One respondent was able to buy an air conditioner, while two others procured water pumps. Refrigerators and freezers are used for both personal and business purposes.

The considerable number of new appliances purchased by residents soon after the project started its 24/7 operation is indicative of the suppressed demand for power in the island that the project was able to satisfy.

New appliances and gadgets purchased. Refrigerators and mobile phones are among the most common appliances and gadgets purchased by residents (photos by Asian Development Bank).

Table 12: Top Ten Appliances Acquired After Project Operation

Appliance	Total Appliances	Year Acquired		Percent Increase in Number of Appliances after the Project
		As of 2018	After 2018 (additional)	
Cellular phone	122	39	83	68
Television	107	56	51	48
Electric fan	125	77	48	38
Rice cooker	76	29	47	62
Washing machine	56	13	43	77
Refrigerator	29	6	23	90
Radio/radio cassette	71	31	40	56
Freezer	16	0	16	100
Electric kettle	20	4	16	80
Computer	11	5	6	54

Source: Survey results.

Service Reliability

The reliability of the service may be gauged through the satisfaction level of consumers and incidence of power interruptions and accidents (if any).

Consumer Satisfaction on Service Reliability

Based on survey results, there is high appreciation for the availability of sufficient electricity service. Ninety-three percent of respondents were satisfied with the availability of 24-hour electricity, citing the longer period of service and ability to use home appliances as their main reason. However, 7% of surveyed respondents also reported unsatisfaction due to unannounced power interruption, which did occur due to systemic and/or management issues.

Incidence of Power Interruption and/or Accidents

Eighty percent of respondents reportedly experienced power interruptions, but only for less than 10 minutes. This happens whenever the power plant transfers its power source from solar to diesel power but the EMS is unable to automatically do the switching (Table 13). Shifting of power source becomes necessary when the power stored in the batteries are no longer sufficient and the diesel generator has to be operated as back-up power source. The incidence of power interruptions in these instances indicates that the shifting of power source from storage battery to diesel is not seamless. This is a critical technical problem that ANTECO engineers must resolve.

Eleven percent of respondents also recalled 1–2-hour power interruptions while 16% experienced 2–6-hour power interruptions. Nevertheless, 98% of respondents indicated that when these interruptions happen, ANTECO restores power usually within the day.

However, an exceptional case happened in 2021. A longer period of interruption occurred due to a malfunction in the software that manages the EMS. ANTECO did not have the source code of the computerized program, as the software being used was proprietary. When this occurred ANTECO needed to communicate with the EMS supplier in the Republic of Korea to fix the problem. This was not easy because the incident happened during the pandemic lockdown. The reliability issue relative to the EMS therefore is a major concern. ANTECO should find a local solution to this problem to avoid reliance on external experts from the Republic of Korea.

No incident of accidents had been reported. However, for some respondents, the intermittent power interruptions cause problems with the use of appliances. This further highlights the issue of power interruptions as a valid operational problem that ANTECO must deal with.

Table 13: Awareness on the Length of Unpredictable Interruption

Length of Unpredictable Interruption	Number	%
Less than 10 minutes	80	54
10–30 minutes	11	7
Over 30 minutes–1 hour	16	11
1–2 hours	16	11
2–6 hours	24	16
Don't know	2	1
Total	**149**	**100**

Source: Survey results.

Level of Satisfaction on Antique Electric Cooperative's Service Delivery

Despite the reported power interruptions, 93% of respondents gave an overwhelming appreciation for the project and the improvement in ANTECO's services, citing the following reasons: (i) longer period of electricity supply, (ii) work has become easier because appliances can be used longer, and (iii) ANTECO is able to do repairs and bring back service in a short period of time.

Comparing ANTECO's performance before and after the project, it is observed that the level of satisfaction has been positively reversed. Whereas 59% of respondents said that they were not satisfied with the electricity service delivered by ANTECO before the project, this number decreased to a negligible 1% after the project. Moreover, the number of those who are satisfied with the project increased from 40% before the project to 63% after the project indicating an increase in level of satisfaction of residents by 23%. Notably also, 36% of respondents said that they are now very satisfied with the improvement of ANTECO's services as opposed to only 1% that were highly satisfied before the project. The level of satisfaction of consumers on ANTECO's service delivery before and after the project are presented in Figure 12. The overall representation of these responses showed that there is a very positive turnaround in the perception of consumers about the performance of ANTECO after the installation of the project.

Figure 12: Level of Satisfaction on Service Delivered by Antique Electric Cooperative

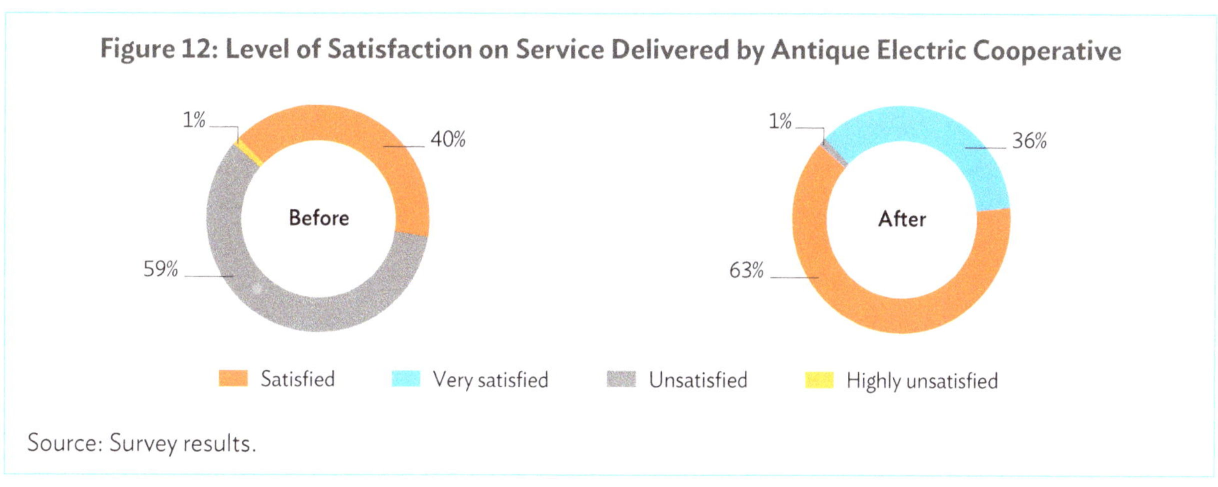

Source: Survey results.

Affordability of Electricity Supply

Increasing Consumption Trend

From the start of the project's operation, there has been a huge improvement in the monthly consumption of consumers. Before the project, average monthly consumption per consumer was only 10.91 kWh in 2017, and 11.20 kWh in 2018, when ANTECO was operating for only 4 hours per day. After the project, monitored data of prepaying consumers showed the average monthly consumption to have increased considerably to 36.40 kWh in 2019 up to as high as 49.90 kWh in 2022. A slight dip in the consumption was experienced in 2021 when there was a total lockdown in the island due to the COVID-19 pandemic. Nevertheless, the increasing trend is seen to be holding as preliminary data on average monthly consumption during the first half of 2023 is already at the level of 47.20 kWh (Figure 13). Demand is expected to further increase in the second half of the year and exceed the 2022 level as people are beginning to travel again and the local tourism economy recovers.

Figure 13: Average Monthly Kilowatt-hour Consumption, 2017–2023

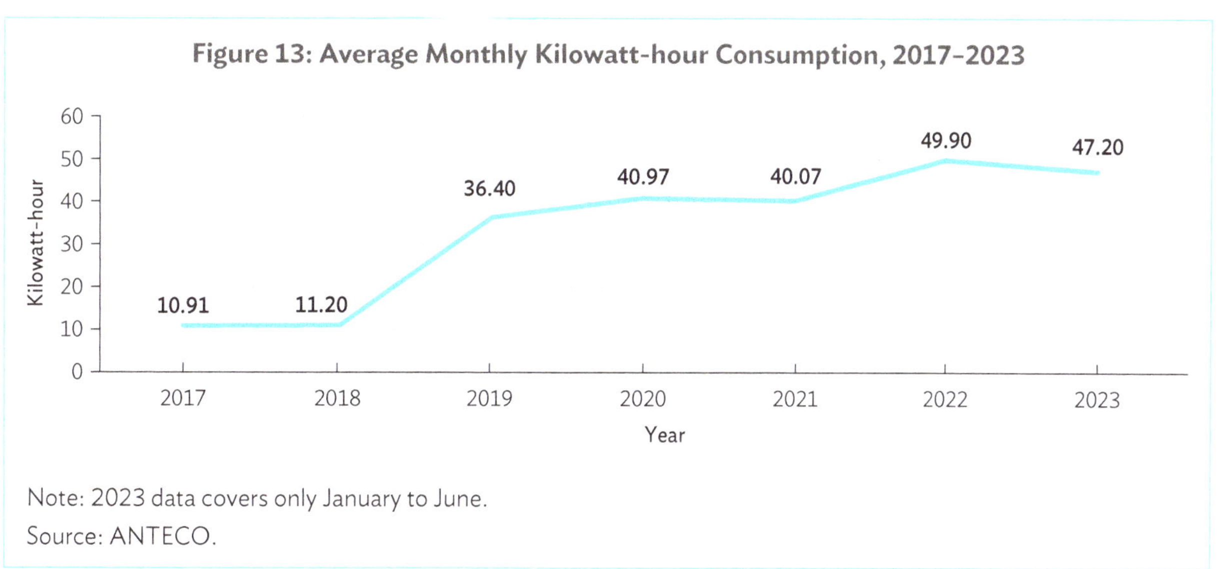

Note: 2023 data covers only January to June.
Source: ANTECO.

Tables 14 and 15 show the number of consumers and their consumption levels in more detail. From the data, it may be noted that over the last 4.5 years (2019–June 2023), the consumption levels of consumers generally increased. This upward movement is seen particularly in consumers using an average of less than 30 kWh of electricity. The number of consumers in this category have sharply declined from 98% pre-project to 57% after the project. Conversely, after the project, 17% of consumers have shifted to the average consumption range of 31–50 kWh from only 2% of consumers in this category before the project. Another 17% have also progressed to the 51–100 kWh average consumption level while 9% are already consuming more than 100 kWh per month. The increase in consumption levels could be attributed to the use of more appliances and entertainment gadgets due to the 24-hour availability of electricity after the project, increased commercial activities, and longer hours of operations as well as lower tariff.

Table 14: Average Monthly Kilowatt-hour Consumption, Before the Project

Average Monthly Consumption, 2017-2018, kWh									
Monthly Consumption Level	2017			2018			Average 2017–2018		
	Consumers		Average kWh Consumption	Consumers		Average kWh Consumption	Consumers		Average kWh Consumption
	N	%		N	%		N	%	
0–30 kWh	127	98	10	119	98	15	123	98	13
31–50 kWh	2	2	37	3	2	40	3	2	39
51–100 kWh	0	0	–	0	0			0	0
>100 kWh	0	0	–	0	0			0	0
Total	**129**	**100**	**10.91**	**122**	**100**	**11.20**	**126**	**100**	**11.05**

kWh = kilowatt-hour, N = number.

Note: Based on a 4-hour/day operation.

Source: ANTECO.

Table 15: Average Monthly Kilowatt-hour Consumption by Prepaid Consumers, After the Project

Prepaid: Average Monthly Consumption, 2019–June 2023 (kWh)									
Monthly Consumption Level	2019			2020			2021		
	Consumers		Average kWh Consumption	Consumers		Average kWh Consumption	Consumers		Average kWh Consumption
	N	%		N	%		N	%	
1–30 kWh	95	67	15	83	59	15	82	57	16
31–50 kWh	24	17	38	26	18	40	26	18	39
51–100 kWh	14	10	68	20	14	72	25	17	70
>100 kWh	9	6	209	13	9	161	11	8	154
Total	**142**	**100**	**36.40**	**142**	**100**	**40.97**	**144**	**100**	**40.07**

continued on next page

Table 15 *continued*

Prepaid: Average Monthly Consumption, 2019–June 2023 (kWh)									
Monthly Consumption Level	2022			2023			Average, 2019–2023		
	Consumers		Average kWh Consumption	Consumers		Average kWh Consumption	Consumers		Average kWh Consumption
	N	%		N	%		N	%	
1–30 kWh	74	53	15	85	51	15	84	57	15
31–50 kWh	24	17	39	22	13	39	24	17	39
51–100 kWh	22	16	72	46	27	51	25	17	67
>100 kWh	19	14	174	15	9	230	13	9	186
Total	**139**	**100**	**49.90**	**168**	**100**	**47.20**	**147**	**100**	**42.78**

kWh = kilowatt-hour, N = number.

Note: Preliminary data (January to June 2023); based on a 24-hour/day operation.

Source: ANTECO.

Blended Tariff

In terms of electricity tariff, ANTECO was able to reduce the price charged to Malalison consumers by ₱20/kWh ($0.40), from ₱30.15 ($0.60) before the project to an average of ₱10.35 ($0.21) per kWh in 2019.[14] This was achieved by using a blended tariff strategy to lower the tariff rates to affordable levels in the island.

To generate the blended tariff, the computation of the generation cost of the Malalison solar PV hybrid mini-grid is included (or blended) into the computation of generation rate for the whole franchise area of ANTECO. The result is a lowered tariff allowing the people of Malalison Island to enjoy electricity service at the same price as those in the mainland. Without the blending of the generation rate into the entirety of the ANTECO franchise, the electricity tariff in the island would have been much higher as the project is still in its capital cost recovery phase. Unlike the NPC-SPUG, ANTECO does not receive UCME subsidy[15] from the government to support its off-grid power generation in the Malalison Island. Thus, in the absence of UCME subsidy, ANTECO applied the blended tariff approach as a strategy which enabled ANTECO to come up with equitable and affordable prices of electricity in the island.

The impact of the blended rate is shown in Figure 14. From 2019-June 2023 the tariff for Malalison averaged ₱33.17 whereas blended tariff averaged only ₱10.76.

[14] Lim Luduvico, presentation slides at the Asia Clean Energy Forum 2019 (unpublished). Conversion rate at ₱50.00 = $1.00.

[15] UCME is a subsidy given in support of NPC-SPUG operations in "missionary areas." Consumers are charged only the subsidized approved generation rate instead of the true cost of power generation. However, electric cooperatives do not have the same privilege to buy down the cost of power generation like NPC-SPUG.

Figure 14: Comparative Tariff, 2019–2023 (Unblended vs. Blended in ₱/kWh)

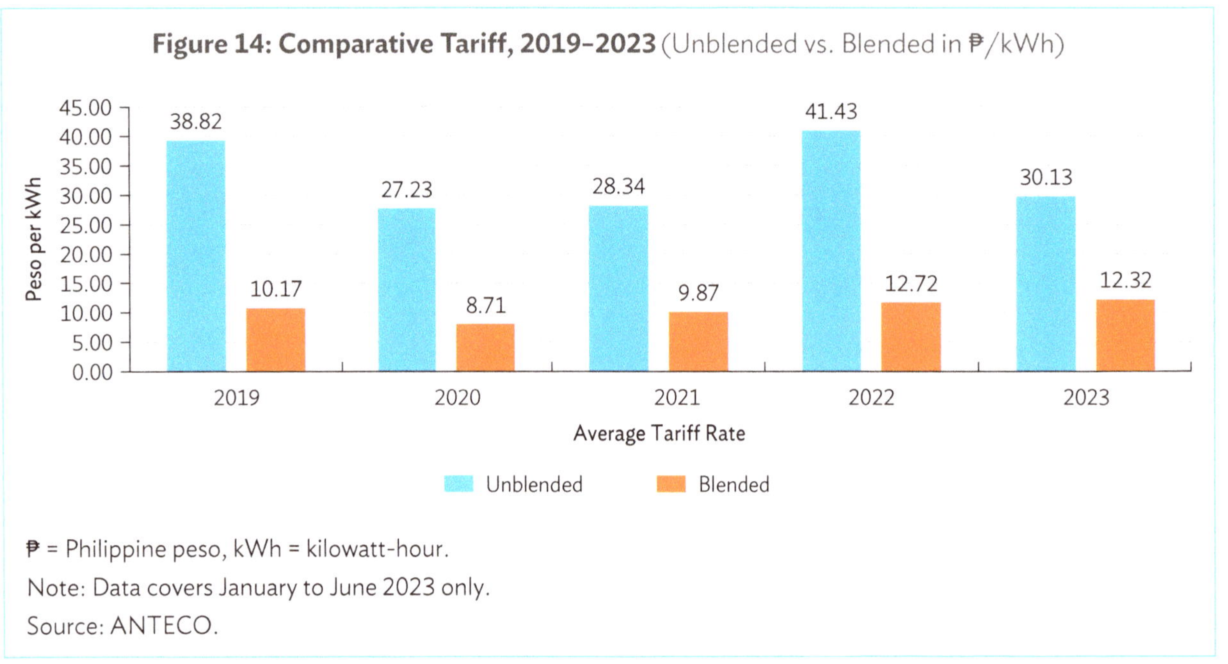

₱ = Philippine peso, kWh = kilowatt-hour.

Note: Data covers January to June 2023 only.

Source: ANTECO.

Subsidy, therefore, remains an important factor in the matter of affordability for off-grid renewable energy projects. Due to the higher first cost for renewables, subsidy is necessary to buy down the true cost of power generation while the capital costs are still being recovered. As demand grows, subsidy may be gradually reduced until such time that the capital costs for the project are fully recovered. The subsidy issue is crucial if DRES projects are to be promoted for deployment in islands not covered by NPC-SPUG.

Tariff Acceptability

The acceptability of the reduced tariffs is reflected in the perception of survey respondents regarding the price they are paying for electricity. Table 16 and Figure 15 show a comparison of the respondents' perception of the electricity price before and after the project. Before the project, 51% of respondents considered the price of electricity as expensive. This number was reduced to 18% after the project. On the other hand, the number of respondents that considered the price as reasonable increased to 57% after the project from only 21% before the project. Moreover, 22% considered the current price as cheap, up from only 7% before the project.

Table 16: Respondents' Perception of the Price on Electricity

Respondent's Perception	Before the Solar PV Project		After the Solar PV Project	
	Number	%	Number	%
Very Expensive	30	**21**	5	3
Expensive	75	**51**	27	18
Reasonable	30	21	85	**57**
Cheap	11	7	32	**22**

PV = photovoltaic.

Source: Survey results.

Figure 15: Respondents' Perception on Price of Electricity

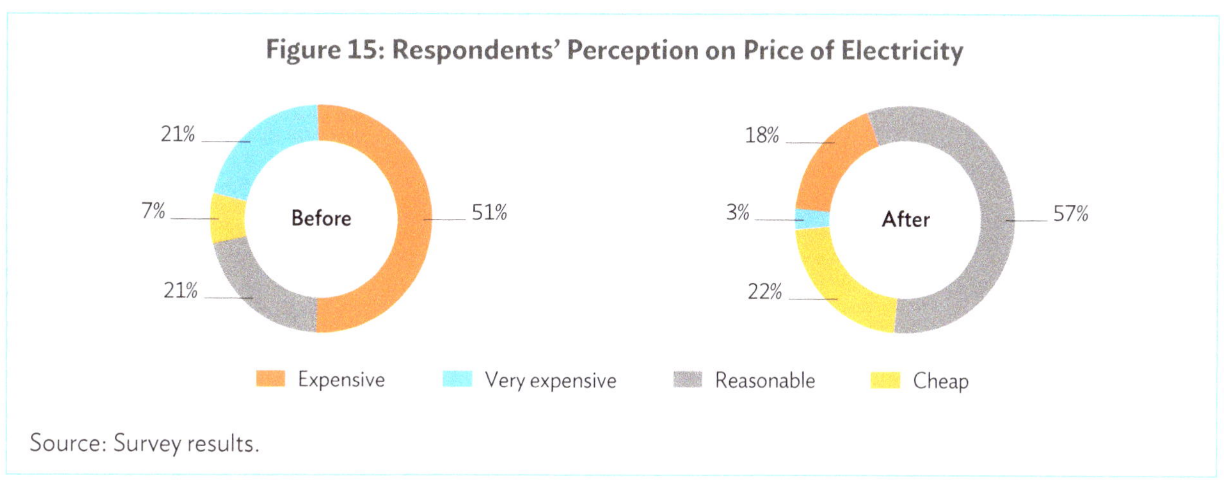

Source: Survey results.

Use of Prepaid Metering System

Survey results also show a high level (97%) satisfaction rate among households on the use of prepaid metering system. As planned, the use of the system brought about energy efficiency to ANTECO and its consumers. The system allowed households, especially daily wage earners and households without fixed incomes to purchase electricity loads in amounts they can afford. This is noteworthy as most household heads are fisherfolk who depend on their daily catch for their livelihood. Even with intermittent cashflow or limited budgets, low-income households can access electricity and manage their usage in a manner that fits their budget. Prepaid metering has also raised awareness about the principles and practice of energy efficiency among Malalison residents. The real-time monitoring of consumption encourages consumers to be more efficient and judicious in their use of electricity. As such, households can avail of electric service without fear of incurring excessive monthly electric bills. Moreover, since the prepaid system is connected to their mobile phones, consumers can conveniently purchase their power in the same way as when they purchase a load for their internet connection. Figure 16 presents the reason why users are satisfied with the prepaid metering system.

Figure 16: Reasons for Satisfaction of Prepaid Metering

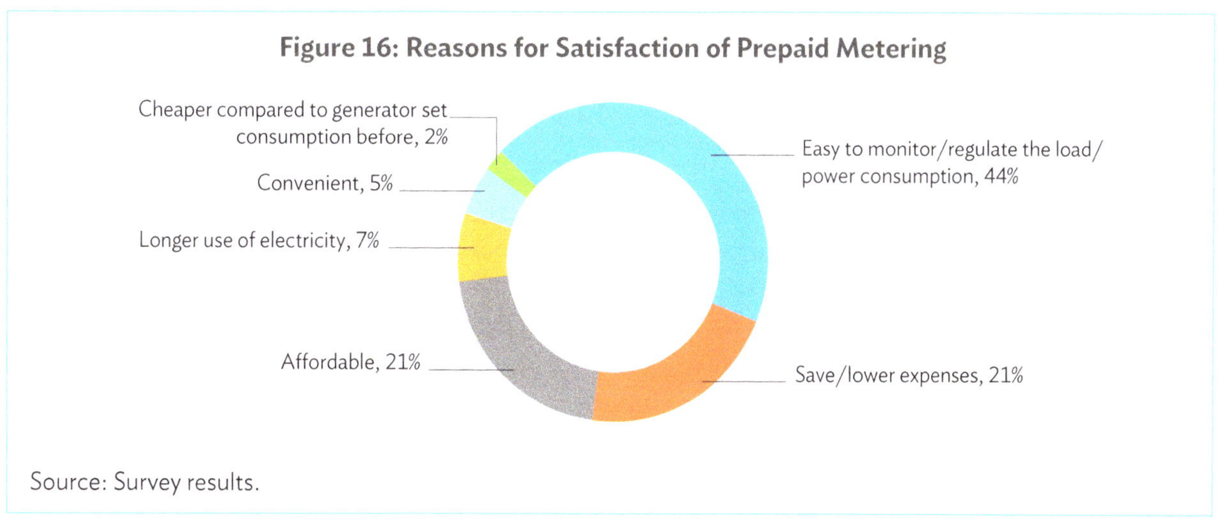

Source: Survey results.

On the part of ANTECO, its operations are now more efficient. The electric cooperative does not have to deploy meter readers, bill collectors, and even disconnection and reconnection engineers, which used to be part of their regular operation activities. Prepaid metering assures ANTECO of 100% collection efficiency with the added benefit of reduced operational cost for billing and collection. Since the system is digital, operations can be monitored remotely as well.

Support for Income Generation

Income Class Classification

The Philippine Statistics Authority defines households by income class based on family income per month as illustrated in Table 17.

Table 17: Philippine Income Class Classification

Income Class	Cluster	Family Income in ₱ per Month	% of Families in Brackets
Low	Poor	Less than 9,100	16.5
	Low Income	9,100 to 18,200	34.4
Middle	Lower Middle	18,200 to 36,400	29.2
	Middle Income	36,400 to 63,700	12.7
	Upper Middle Class (Phils)	63,700 to 109,200	4.9
High	Upper Class	109,200 to 182,000	1.7
	Rich	Above 182,000	0.6

Source: Philippine Statistics Authority. Derived from A. Divina. *Middle Class in the Philippines: Defining Income Class and Ranges*. Digido.

According to this classification, majority of the households in Malalison Island fall under the low-income category and within the poor and/or low-income cluster. Table 18 shows the household monthly income in Malalison Island as per information given by household heads during the survey. Majority of the households (81%) have an income of within the ₱10,000 and below range in 2020, and considered as poor to low income. Around 12% are earning between ₱10,001–₱20,000 per month, classified as low income. Only about 3% of households may be considered lower middle, another 3% in the middle-income bracket and 1% classified in the upper middle class. No one falls under the upper class category.

Meanwhile in the latest 2021 Family Income and Expenditure Survey done by the Philippine Statistics Authority, the regional average annual income for household in the province of Antique is ₱244,530 and average annual expenditure is ₱163,760.[16] Again, comparing this with the 2020 household income in Malalison (Table 18), it can be concluded that 93% of the households i.e., those earning monthly income of ₱20,000 and below, fall short of the average provincial family income level. Hence, whatever support for income generation that the project can deliver will go a long way to improving the livelihood and quality of life of the people in Malalison Island.

16 Philippine Statistics Authority. 2022. Highlights of the Preliminary Results of the 2021 Annual Family Income and Expenditure Survey. Press release. 15 August.

Table 18: Malalison Household Monthly Income

Income Range (₱)	2018		2019		2020	
	Number	%	Number	%	Number	%
10,000 and below	94	63	89	60	120	81
10,001 to 20,000	35	23	40	27	18	12
20,001 to 30,000	10	7	10	7	5	3
30,001 to 40,000	4	3	2	1	1	1
40,001 to 50,000	3	2	1	1	3	2
50,001 to 60,000			2	1		
60,001 to 70,000			1	1	1	1
70,001 to 80,000	1	1	2	1	1	1
80,001 and above	2	1	2	1		
Total	**149**	**100**	**149**	**100**	**149**	**100**

Source: Survey results.

New Income-Generating Activities

It is not possible to directly attribute the existence of new business start-ups and expansions to the project, as this is a "before–after" (not "with–without") comparison. However, given the magnitude of opportunities that are now open due to the availability of sufficient supply of 24-hour electricity, the expectation is that the majority of changes in the economic landscape will be due to the project.

From the survey, it is notable that a number of households have taken advantage of the opportunities for new income generation since the start of the project's operation. Household respondents reported starting 61 new businesses after the project extended ANTECO's service to 24-hours. Being a travel destination, the operation of homestay and resort facilities were the most popular and lucrative. Haida Doroteo, an enterprising woman, operates a homestay business to augment her family's income from fishing (Box). She appreciates the solar-powered electricity on the island. Other new income-generating activities include convenience stores and souvenir shops, food stalls, and cooking services as well as sale of frozen foods. With the acquisition of refrigeration appliances, women and housewives were able to engage in supplemental income-generating activities, such as sale of ice, cold beverages, and iced refreshments. Table 19 presents the number and types of new business activities and the estimates of average income these new businesses will be able to generate. Homestay businesses, stores and souvenir shops, food stalls, and businesses related to food are mainly run by women, while boat operations and tour-guiding are done by the men.

For the 61 households that ventured into new income-generating activities, the additional income would likely enable them to overcome the low-income threshold and move up to middle-income classification (Table 17).

Box: Haida's Homestay

Haida Doroteo is a 53-year old enterprising woman who calls Malalison Island home. She set up her business to have an alternative source of income to provide for her family. During the coronavirus disease (COVID-19) pandemic, life was harder on the island without the tourists. Though they rely on fishing for their livelihood, this does not provide as much as tourism. She says the presence of electricity and solar power have made things more comfortable for them on the island because they can properly store food and keep their place cool at all times of the day. For her guests, she recommends swimming, hiking, snorkeling, and boating while on the island.

Discovering Malalison Island. Developing the tourism sector in Malalison Island (photos from the Municipality of Antique website. https://culasiantique.gov.ph/malalison-island-2/).

Source: Haida Dorotea's Homestay. https://www.banwacommunity.com/homestay/haida-homestay/.

Table 19: Business Activities Initiated and Estimated Additional Income Generated

Business Venture	Number	%	Estimate of Average Additional Income (₱)	
			Monthly	Annually
Commercial	**60**			
Homestay/resort	20	33	19,175.00	230,100.00
Stores and souvenir shops	17	28	5,106.00	61,272.00
Boat operations	11	18	4,364.00	52,368.00
Food stall/cooking services	9	15	4,133.00	49,596.00
Frozen foods	2	3	350.00	4,200.00
Site guiding	1	2	1,500.00	18,000.00
Agricultural (Fish Culture)	**1**		1,500.00	18,000.00
Total	**61**			

Source: Survey results.

Entrepreneurial housewives. Photos taken during the household survey show some of the women-led income generating activities in the island (photos by Asian Development Bank).

In terms of consumption pattern, an increasing trend is seen in the per capita electricity consumption in Malalison. Compared to the mainland Culasi municipality to which Barangay Malalison belongs, Malalison's per capita consumption level is on the average 63% that of the mainland. If the increasing trend continues due to improved business environment and more people undertaking new income-generating activities, the consumption levels can most likely improve further to approximate that of the mainland. This would also improve the project's sustainability in the long run (Figure 17).

Figure 17: Comparative per Capita Consumption (kilowatt-hour, Culasi vs. Malalison)

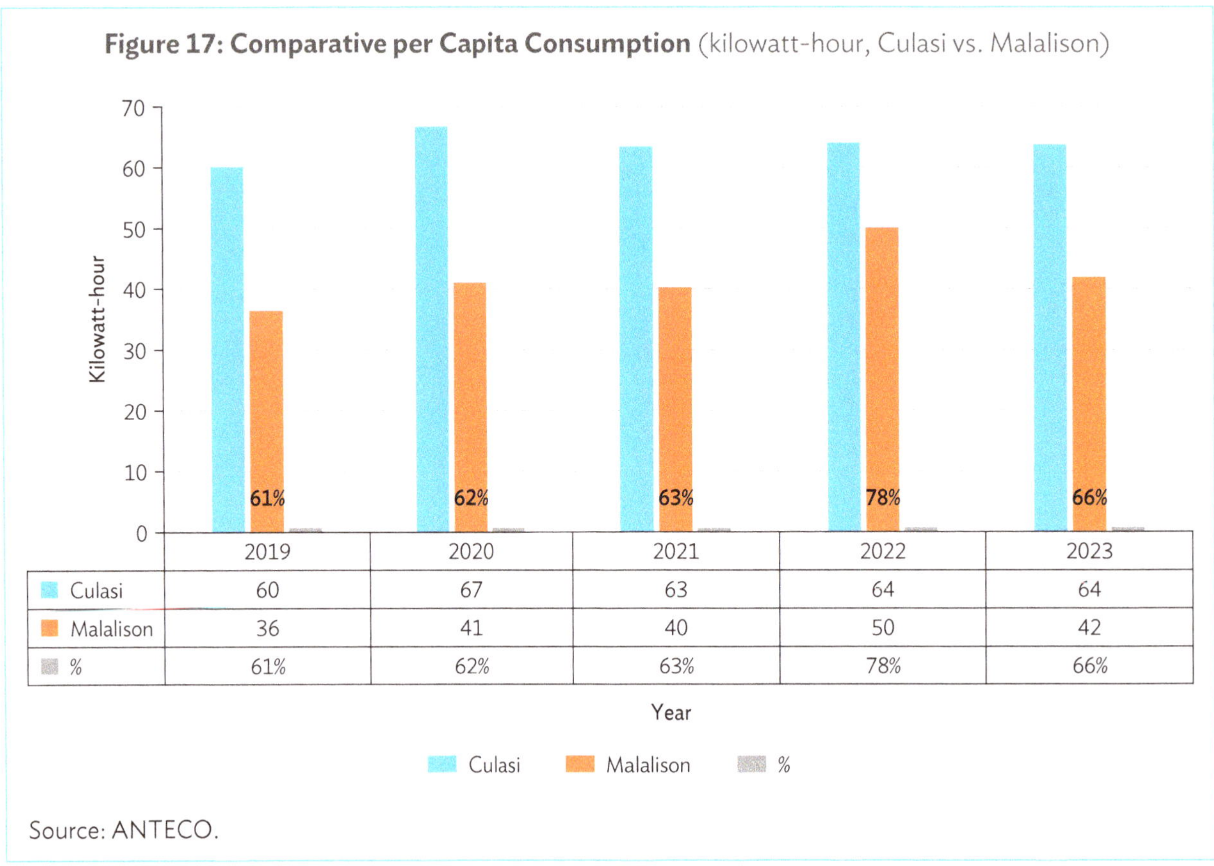

	2019	2020	2021	2022	2023
Culasi	60	67	63	64	64
Malalison	36	41	40	50	42
%	61%	62%	63%	78%	66%

Source: ANTECO.

Thirty percent of respondents attested that their income increased soon after the project became operational. According to them, these increases in income were the result of more tourist visits to the island and increased sales because of longer hours of operations and additional product offerings.

Respondents with commercial operations revealed that their hours of operation significantly increased after the project's installation. Homestay and resort businesses extended their operating hours to 24 hours, from 8 hours per day before the project. Similarly, the hours of operations of convenience stores increased from 12 hours to 16 hours daily. In addition, business operations improved as these establishments were able to offer other services. Homestays and resorts were able to provide air-conditioned rooms to visitors. Table 20 shows the perceived improvements in business operations by respondents.

More than half of survey respondents appeared optimistic about the future in terms of opportunities for investments. When asked whether they had plans of undertaking new businesses, 55% responded in the positive. The new business being contemplated include: sale of frozen food; minimart or *sari-sari* (convenience) store; sale of ice; cooking service, food and refreshment stalls; and homestay or resort. Again, most of the income-generating activities contemplated are in the food business and those intending to serve the island's tourism industry (Table 21).

Table 20: Improvements in the Business Operations

Perceived Improvement	Homestay	Resort	Convenience Store	Number	%
Longer business hours	4	2	6	12	27
Acquired appliances for lighting, ventilation, refrigeration, and cooling	4	2	1	7	16
Increased income from services	1		5	6	14
Acquired internet connection	4	2		6	14
Offers mobile loading			5	5	11
Offers cold products and drinks, food catering			4	4	9
Savings in electricity cost with efficient use			4	4	9
Total Respondents	**13**	**6**	**25**	**44**	**100**

Note: Multiple answers were gathered from survey results.

Source: Survey results.

Table 21: Planned New Investments

New Business Being Contemplated	Number	%
Yes	**82**	**55**
Selling frozen foods	26	17
Sari-sari store	25	17
Selling ice	12	8
Food stall/cooking services	6	4
Homestay	6	4
No plan	**65**	**45**
Total	**147**	**100**

Note: Multiple answers were gathered from survey results.

Source: Survey results.

Personal and Household Convenience

This section focuses on the effects of the pilot project on their living condition as perceived by household users. Table 22 presents the responses of surveyed household heads on the positive effects of having 24-hour electricity in the island. Topping the list is the convenience of having well-lighted homes and surroundings, followed by being more productive as household chores become easier to manage because of having useful appliances and being able to use them anytime.

Table 22: Positive Impacts of the Pilot Project on the Household Living Conditions

Perceived Positive Impacts	Number	%
Well-lighted home and surroundings	88	59
Eases household chores	70	47
Feel happy and comfortable, and convenience in daily living	59	37
Feel safe and secure at night	53	36
Well-ventilated room	25	17
Increased livelihood income and small businesses	21	14
Additional appliances were acquired which can be used anytime	20	13
Students can study at night	19	13
Clean environment, less pollution	12	8
Access to information	10	7
Total Respondents	**149**	**100**

Note: Multiple answers were gathered from survey results.

Source: Survey results.

The 24-hour electricity that the pilot project was able to provide has changed the course of daily household activities. In Table 23, the changes in the time devoted to housework, entertainment, business, and other activities are shown. Time devoted to housework has generally shortened, giving household members more time to do productive work such as preparing for the next day's business activities. Business hours have been extended and some respondents were able to enjoy longer home entertainment and do outside activities, including youth education.

Table 23: Changes in Duration of Doing Activities with 24-hour Electric Services

Duration	Number	%
Shorter	**69**	**46**
Housework (have time for other productive activities)	65	94
Home entertainment	18	26
Business (preparation time)	3	4
Longer	**31**	**21**
Home entertainment	27	87
Housework hours (can do more work, even at night)	21	68
Business hours	11	36
Youth education	2	7
Outside activities	1	3
Same	**49**	**33**

Note: Multiple answers gathered from survey results.

Source: Survey results.

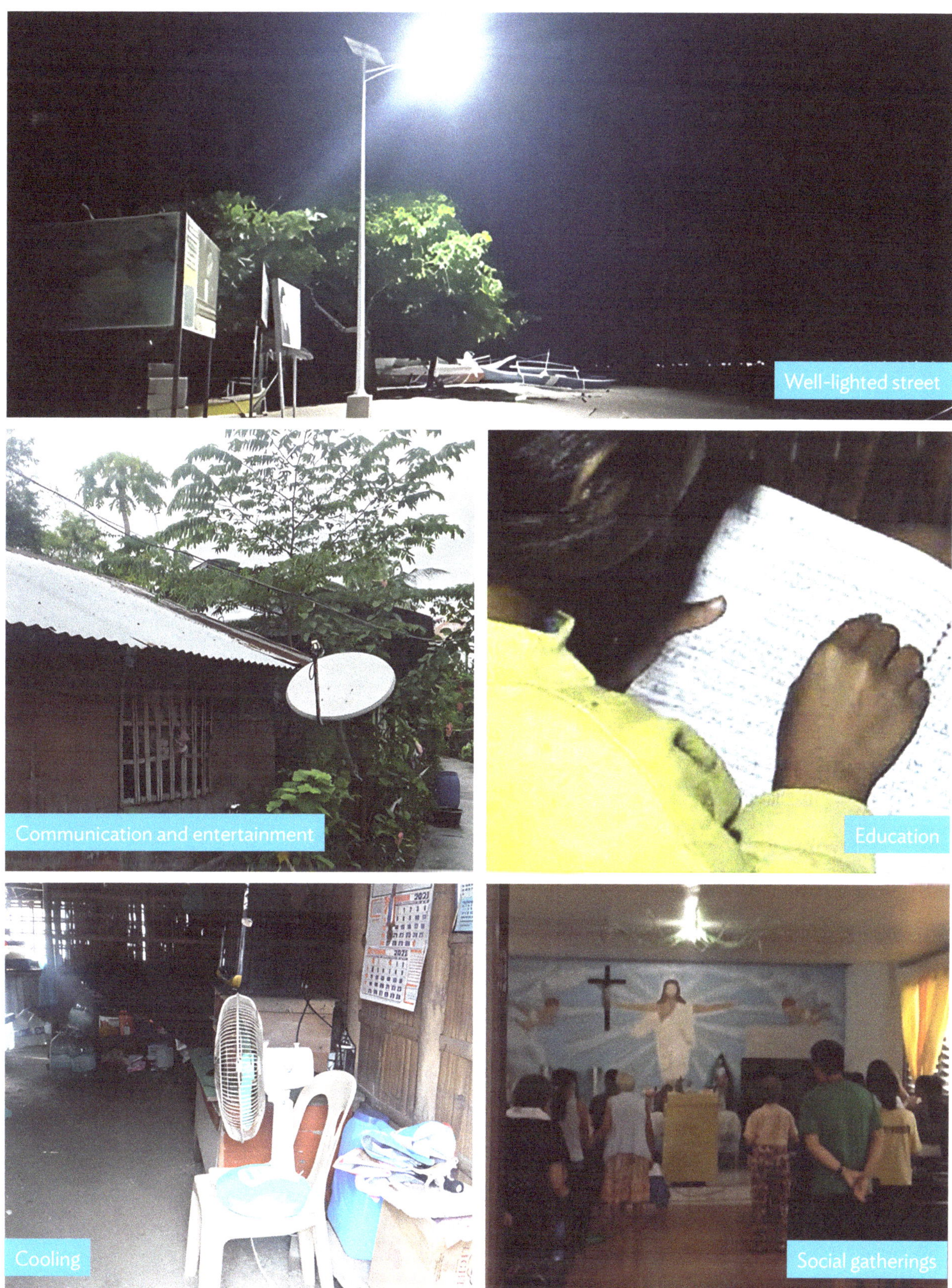

Benefits from the use of electricity. People of Malalison Island now enjoy the benefits due to the 24-hour service provided by the project (photos by Asian Development Bank).

Project Impact on the Community

To assess the impact on the community, respondents were asked about their opinion on the changes they observed in their community. Interviews were also conducted with (i) teachers on the impact on education, (ii) community organization members on their activities, and (iii) ANTECO manager on its service to the community. The findings are presented in this section.

Household Perception of Project Impact on the Community

The top five positive changes in the community (barangay) that generated most impact are

(i) improvements in education (children studying longer and at night),

(ii) environment cleanliness,

(iii) safety and security,

(iv) children's health (non-exposure to kerosene fumes), and

(v) improved business climate.

More comfortable classrooms and better teaching materials were also noted. Other positive changes include improved barangay services in the health center and ability to organize barangay activities. Respondents also mentioned that they have improved social life as they are now able to hold social activities at night. The list of positive changes in the community based on the survey is shown in Table 24.

Table 24: Observed Positive Changes in the Community Due to 24-hour Electric Service

Observed Positive Changes	Number	%
Children can study longer and can study at night	128	86
The barangay is clean and green	114	77
It is safer at night, with well-lighted streets and beachfront	107	72
There are more shops, more businesses	95	64
Children are no longer exposed to kerosene fumes	93	62
Services in the barangay have improved and/or new infrastructure added	90	60
Chapel services can be held even at night	87	58
Less youth migration to cities; residents who left have come back	80	54
Classrooms are more comfortable and more teaching materials, such as computers are available	74	50
New community organizations have been established	65	44
Social life in the barangay has improved; activities can be held at night	62	42
Number of diseases and health-related cases decreased; the capability of the health center to deal with health-related cases has improved	43	29

Note: Multiple answers gathered from survey results.

Source: Survey results.

Barangay Services

With sufficient power day and night, the social services offered by the barangay improved significantly. The improvement most noticed by barangay residents are the access to information and improved security due to street lights that the barangay have put up. Education and health service likewise greatly improved especially the availability of vaccine during the COVID-19 pandemic. Table 25 shows the major social services that have been introduced and/or improved by the local government unit (barangay).

Table 25: Barangay Social Services Introduced or Improved After the Project

Barangay Services	Number	%
Availability of vaccine in health center	116	78
Improved educational facilities	119	80
Use of technological devices and other multimedia platforms in conducting classes, parent–teacher meetings and other activities		
Use of computer and printer for learning materials		
Better ventilation and lighting		
Construction of additional classrooms		
Installation of solar streetlights		
Access to information	140	94
Improved barangay security	139	93

Note: Multiple answers gathered from survey results.
Source: Survey results.

Education

Barangay Malalison has an elementary school only. To get information on the project's impact on education, schoolteachers were interviewed to confirm the perception of household heads about the impact of the project on their children's education. The teachers' views centered on the significant improvements in the conduct of classroom instructions: their ability to use multimedia-aided instructions, such as the use of videos from YouTube and other online sources, and references in class, which have boosted the pupils' interest in learning.

According to the teachers, the availability of online learning materials allowed them to use varied teaching strategies for different types of learners as compared before when teachers would conduct classes simply using the traditional lecture method. The availability of computers and printers paved the way for increased use of printed learning resources and materials they can distribute to students, especially during the pandemic when face-to-face teaching was not allowed. Lights and electric fans improved the ambience inside the classroom, providing more comfort and better ventilation. With better facilities and instructional materials, there were also observed positive changes in the behavior, attitude, attendance, and performance of students. These observations are shown in Table 26.

Table 26: Improvements in the Behavior, Attitude, Attendance, and Performance of Pupils

Behavior
- Pupils show enthusiasm to watch downloaded videos from YouTube
- Students show interest in class

Attitude
- Learners are more receptive toward the lesson
- There is more retention

Attendance
- Absences were minimized

Performance
- Most of the learners performed better academically
- Improved listening and comprehension skills
- Improved reading skills

Note: Multiple answers gathered from survey results.
Source: Survey results.

People Empowerment

Three local organizations in the island: Malalison Island Community-Based Eco-Tourism (MICBETO), Department of Agriculture Fisher Folks Association, and Malalison Island Children's Choir (MICC) reflect the main industry and interest of the people—fishing, tourism, and education.[17]

From the year the project operated, membership of all three organizations showed an increasing trend. The three organizations also reported improvements in their activities and new activities being undertaken by their members (Tables 27 and 28). Notably, the organizations now can undertake important activities, such as meetings and even income-generating activities at night as well as sale of frozen goods and cold drinks during special events. Even the children's choir, which helps in welcoming tourists in the island shared being able to use loudspeakers for their choral presentations and hold their voice and choir rehearsals at night.

Table 27: Membership in Organizations in Malalison Island

Name of Organization	2018	2019	2020
Malalison Island Community-Based Eco-Tourism	300	302	307
Department of Agriculture Fisher Folks Association	160	181	203
Malalison Island Children's Choir	37	46	66

Source: Survey results.

[17] MICBETO and MICC are known locally as "Mararison Island Community-based Eco-Tourism" and "Mararison Island Children's Choir."

Table 28: New Activities and Improvements in Organization Activities After the Project

Organization	Activities that Improved	New Activities Undertaken
Malalison Island Community-Based Eco-Tourism (MICBETO)	• Massage services can be made available at night • Tour-guiding can extend at night due to the presence of streetlights along the way	• Selling of frozen foods, ice candy, massage, souvenirs, ice cream, cold buko juice
Department of Agriculture Fisher Folks Association	• Improved presentation during meetings • Meeting can be conducted at night	• Preservation of fish • Crown-of-thorns (*damang damang*) removal[a]
Malalison Island Children's Choir (MICC)	• Voice coaching and lessons • Practice of choreography for the chorale's presentation • Use of speakers for choral presentations	• Rehearsals at night

[a] Crown-of-thorns starfish, *Acanthaster planci*, is a large starfish that preys upon hard, or stony, coral polyps and threatens the life of corals.

Source: Survey results.

Possible Negative Behavioral Changes

Twenty-four-hour energy access can also bring about some behavioral changes to community life, which may have some negative consequences. Heads of households voiced some concerns and observations they feel could lead to negative behavior most especially among children and youth. Among the most significant of these are the observed excessive use of gadgets, which some household heads fear could lead to addiction. With the ability to stay up late at night, community members may also be encouraged to excesses, such as staying out late for parties, drinking sessions, and other vices. This could lead to less time for studies, as students get distracted by social media, less family bonding, and the deterioration of relationships among household and community members (Table 29).

Table 29: Households' Perception on Negative Behavioral Changes in the Community

Changes Considered as Negative	Number	%
Noisy environment due to late night gatherings	63	42
Excessive use of gadgets	53	36
Family members and residents staying up late at night	41	28
People engage in drinking liquor and other vices	40	27
Overuse of energy	14	9
Less time to study because students get distracted	4	3
Less bonding time with family	4	3

Note: Multiple answers gathered from survey results.

Source: Survey results.

Antique Electric Cooperative Operations

From the point of view of ANTECO management, there are noteworthy improvements in the services offered by the electric cooperative particularly in providing sufficient, reliable, 24/7, clean, and environmentally friendly electricity. The installation of the solar PV hybrid mini-grid project is a testament to ANTECO's commitment to find a clean and sustainable solution toward providing energy access to small islands within the electric cooperative's franchise coverage. ANTECO has other islands that can benefit from the experience of the Malalison Island solar PV hybrid pilot project.

ANTECO's use of prepaid metering system also benefited both the electric cooperative and their consumers. For ANTECO, the prepaid system reduced its expenses for personnel traveling to the island for meter reading, bill preparation, and collection since the system can automatically process the households' electric consumption virtually. There is also assurance of 100% collection efficiency for the electric cooperative, as electricity consumed by users are paid for in advance. As consumers become more conscious and watchful on their use of electricity, they can regulate and be more judicious in the use of their electricity consumption. Residents are also vigilant on the occurrence of outages and this provides pressure on ANTECO to perform better and deal with the operational challenges being experienced.

Technically, ANTECO encountered two main challenges in its operations. These included: (i) troubleshooting the malfunction in the system, and (ii) increase in demand due to influx of tourists and increased demand from additional appliances purchased by residents. With increasing consumption, ANTECO sees the need to increase the capacity of the system soon to ensure that it will not experience overloading. This need to increase capacity has been anticipated as the pilot project was originally designed to serve only the projected demand of 200 households in 5 years. Today, the solar PV hybrid plant is already on its fifth year of operation.

The project would at times experience technical malfunctions in the operation of the EMS for the solar batteries and automatic switching to diesel source of power. This happens when there is loss of communication, and the EMS software encounters a problem. Since the digital software is proprietary, ANTECO does not have the source code to make the necessary programming changes in the software. When this happens, ANTECO engineers have to contact and get assistance virtually from the Republic of Korea. Due to the COVID-19 lockdown, the Korean supplier was unable to complete its face-to-face training and instead updates and tutorials on system restoration are being provided, but done virtually. This is a major problem that ANTECO acknowledges and is finding ways to resolve. There appears to be a need to localize the supply for equipment like the EMS and improve the operation and management capacity of ANTECO. This is one important lesson that became evident during the operation of the pilot project.

ANTECO management confirms both the positive and negative impacts of the solar PV hybrid project as observed by the island residents themselves. According to ANTECO, notable of the benefits to the community are (i) increased livelihood and business establishments, (ii) improved infrastructure with utilization of construction equipment, (iii) upliftment in standard of living and fish catch preservation by fisherfolk, (iv) improved students' learning and interest, and (v) access to information through the internet. On the other hand, noise pollution has been noted due to use of sound systems, more residents staying out late at night, and addiction to gadgets.

Project Impact on the Environment

When 24/7 electricity service was achieved using the solar PV hybrid system with battery storage, there was less usage of fossil fuel. This means that potential greenhouse gas (GHG) emissions from diesel, if it were used 24/7 is largely displaced by the solar PV system. As shown in Figure 18, the project was able to generate on the average 77% of its power from the island's solar resources. This average solar generation is 3% lower than the 80% target solar share as per the project's technical design. The reason is the weather condition and duration of sunlight hours which fluctuates. Also, solar power generated in 2021 was only 65% of target due to technical problems encountered by the pilot plant's EMS. However, the solar penetration rate was at 82% in 2020 and 81% in 2022, exceeding the target. The solar share in the first half of 2023 is also looking promising at 81%.

Figure 18: Solar Power Share to Total Generation (2019–June 2023)

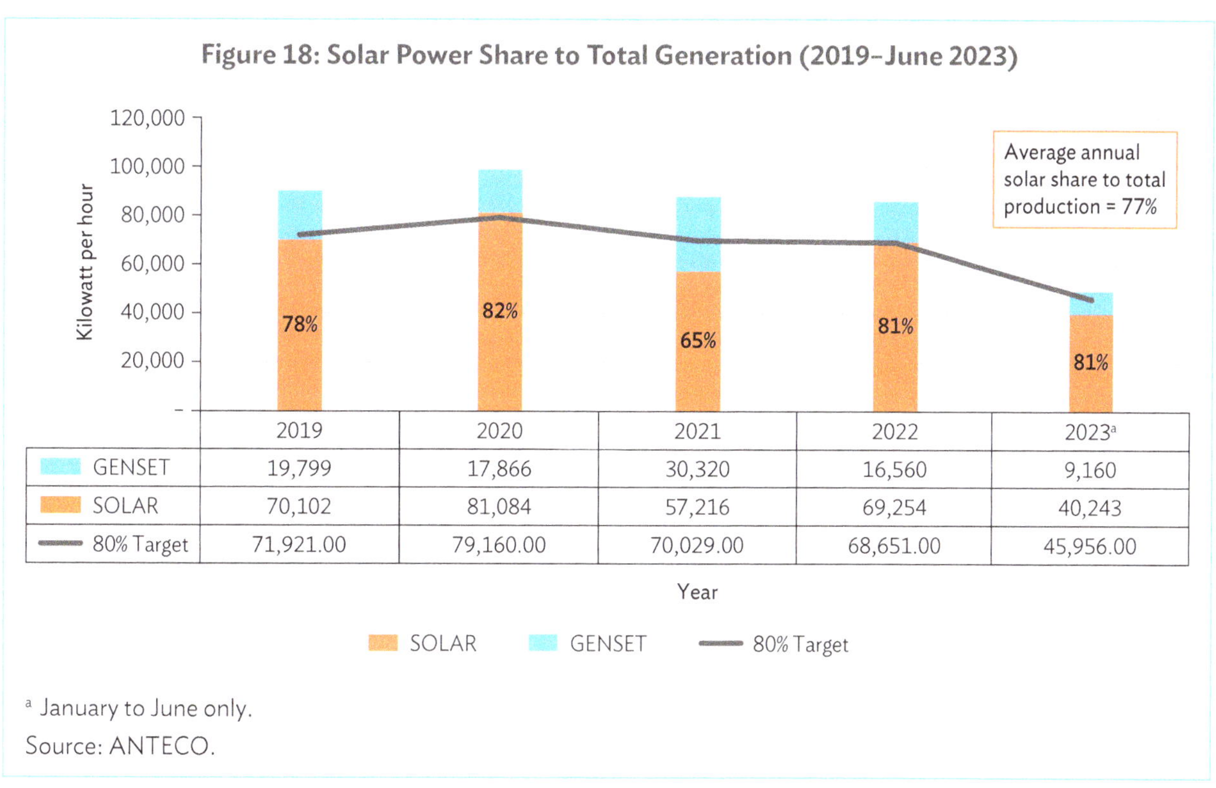

	2019	2020	2021	2022	2023[a]
GENSET	19,799	17,866	30,320	16,560	9,160
SOLAR	70,102	81,084	57,216	69,254	40,243
80% Target	71,921.00	79,160.00	70,029.00	68,651.00	45,956.00

[a] January to June only.

Source: ANTECO.

Over the plant's 4.5 years of operation, it was able to generate 317,899 kWh from solar resource. At the equivalent rate of 1.2 kilograms CO_2/kWh of GHG emissions displaced, if solar is used instead of diesel fuel, the project is estimated to have mitigated GHG emissions corresponding to about 381.48 tons CO_2 equivalent or 84.77 tons CO_2 equivalent annually.

Surveyed respondents view the project's impact on the environment in a positive light. As per survey findings, 100% of respondents believe the project has contributed to a cleaner environment in the island. Respondents also manifested awareness about the pollutive characteristic of diesel fuel and are bothered by the noise when the genset is running. Reduction of pollution from use of kerosene lamps for lighting and fuel wood for cooking are also cited as among the benefits to the community.

This has resulted in positive impact on women and children who are most exposed to fumes from kerosene lamps and indoor air pollution when cooking with fuelwood. Table 30 presents the household heads' perception on the project's environmental benefits.

Table 30: Households' Perception on the Project's Contribution for a Cleaner Environment

Environmental Contribution	Number	%
Yes	**149**	**100**
Less pollution from the diesel generator set	141	95
No more noise from the generator set	118	79
No more pollution from the kerosene lamp for lighting and fuel wood for cooking	115	77

Note: Multiple answers gathered from survey results.

Source: Survey results.

Impact of the Coronavirus Disease

The COVID-19 pandemic wrought tremendous impact on lives of the people of Malalison Island. Foremost, the island went into a total lockdown which lasted for 2 years, totally shutting the island from its tourism activities. Consequently, 100% of households and business establishments experienced loss of income through closure and unemployment for those who depended on servicing tourist activities such as tours and logistics (boat rentals, etc.). Limited movement also affected fishing activities. As can be noted in Table 18 the number of households with income level of ₱10,000 a month and below increased to 81% in 2021 during the pandemic from 63% in 2019—indicating that more people were earning less than before.

Likewise, monthly electricity consumption for consumers that use over 100 kWh per month of electricity (assumed to be those with commercial businesses) declined during the pandemic from the average of 209 kWhs in 2019 to 161 kWhs and 154 kWhs in 2020 and 2021, respectively (Table 15). However, recovery was observed beginning 2022 and consumption could be back to the pre-pandemic level by 2023. This trend is expected to further increase when tourism activities fully normalize and business conditions improve.

Relative to the power supply condition, the use of prepayment metering system sets the Malalison Island experience apart from other areas during the COVID-19 lockdown. With the prepaid system, the residents continued to use electricity without generating debts for accumulated bills over the duration of the lockdowns. ANTECO's cashflow was also not affected by increased receivables due to people not paying their bills. In the mainland, ANTECO had to implement a moratorium on payment of electricity bills for many months. But in Malalison Island, there was no need for this as consumers were purchasing electricity on a prepaid basis. Households simply adjusted and became more efficient in their use of electricity to cope with the existing situation. Moreover, the solar PV technology provided resilience to ANTECO's operations in the island during the COVID-19 lockdown. Unlike in other islands that were dependent on diesel fuel for power generation, Malalison Island did not experience any power supply insufficiency (except when the EMS had a problem) due to the high cost and difficulty in transporting diesel fuel to the island during the transportation lockdowns.

5 SUMMARY OF ASSESSMENT FINDINGS

The Malalison Island Solar PV Hybrid Project achieved positive outcomes in all fronts of the energy trilemma: energy security, energy equity, and environmental sustainability. Additionally, from the social welfare angle, the project was able to enhance basic services, improved the general living standards, and empowered the people to engage in their own businesses and work toward a better future. All these were made possible through the partnership efforts of the major stakeholders of the project: ANTECO and its private sector partner, OREEi, the LGU of Culasi and barangay unit of Malalison, and the people of Malalison Island with the support of ADB and NEA. The project highlights the value of partnerships and collaboration in the introduction of innovative and strategic solutions to achieve shared development goals.

Energy Security

The project had a positive impact on achieving the United Nations Sustainable Development Goal 7, that of addressing energy access. It enhanced the power generation capacity and extended the hours of ANTECO's service in Malalison Island from 4 hours daily to 24 hours. Moreover, with the use of solar energy and battery energy storage system, energy availability is assured even during months when transport of diesel fuel to the island may be affected by inclement weather conditions or during public emergency situations, like in the case of the COVID-19 pandemic.

As of June 2023, 24/7 service was made available to 188 grid-connected consumers, including 95% of the reported number of households in the island plus six commercial establishments and communal facilities. Aside from household lighting, the sufficiency and reliability of power supply from the solar PV hybrid project enabled consumers to purchase electrical appliances, such as refrigerators, rice cookers, washing machines, etc., which they now use to lighten their household chores, enhance their livelihood, and improve their living conditions. Children likewise were able to study at night and teachers obtained access to educational tools, such as computers, internet, etc., which greatly enhanced their teaching methods and strategies. This largely improved the school's educational services, as teachers became effective in teaching varied types of learners. Despite the lockdown during the pandemic, schoolchildren in Malalison Island were able to finish their academic curriculum for the school years 2020–2021 using modular approach, which was developed by the Department of Education to deal with the lockdowns.

Majority of household heads surveyed (93%) were satisfied with the electricity services they receive. There is also overwhelming appreciation of the project by 100% of respondents citing the longer electricity supply availability and ease of housework because appliances can be used longer. However, a small number of respondents have remained unsatisfied because of voltage fluctuations and interruptions which they experienced, although ANTECO was reportedly able to fix or restore service in a short period of time. This is an operational issue that ANTECO must deal with and improve on.

Energy Equity

The project has likewise generated positive impacts on energy equity and inclusivity. With the ₱20/kWh reduction in tariff, the price of electricity has become more affordable. By using the blended rate calculation approach, ANTECO was able to lower the electricity tariff to the same level to that of the mainland. This is a significant development as this has provided electricity tariff equity that the people in Malalison Island have longed for in a long while. About 79% of respondents perceived the price of electricity now to be reasonable and/or cheap. With the project, there has been an observed increase in average monthly consumption per consumer from 11 kWh in 2017–2018 (before the project) to 42.78 kWh during 2019–2023 (after the project). This is largely due to the availability of 24/7 electricity service which allowed the utilization of additional appliances, longer hours of commercial operations, and the lowered electricity tariff.

The use of blended tariff to lower the electricity tariff in Malalison Island is important because ANTECO's operation in the island is not the same as islands under the NPC-SPUG that receive UCME subsidy from the government. Without blending of the tariff with the mainland, electricity tariff in the island would be much higher, until such time that the investment cost of the project is fully recovered.

The prepayment metering system also generated positive feedback from consumers. Households can purchase their electricity according to their affordability and available cashflow. There is no monthly electric bill shock because the prepaid metering allows consumers to monitor their consumption, so they were able to carry out energy efficiency measures to conserve precious funds. This feature fostered inclusivity and encouraged energy efficiency among users. On the other hand, the use of the prepayment system assured ANTECO 100% collection efficiency, while generating savings from nondeployment of people for meter reading, billing, and collection of monthly payments from consumers. Likewise, with the prepaid system, ANTECO also avoided the cost for power disconnection and reconnection in case of nonpayment by consumers.

The project supported the livelihood and income-generation activities mainly due to the provision of sufficient power supply on a 24/7 basis. According to survey results, 61 new businesses were initiated as soon as the project became operational. Most of these businesses are tourism-related, as the island has been included as one of the tourist destinations in the Western Visayas region. Twenty of these new businesses are homestay or resort facilities that offer overnight accommodation to tourists. These homestay or resort businesses reportedly generated about ₱19,000 a month of income. Other businesses, such as convenience stores, eateries, and tour-guiding and boat services also generated about ₱5,000 of additional income, monthly. The additional income were derived from longer hours of operation and expanded or improved services, such as cooling (refrigeration, air-conditioning), internet, convenience stores, food services, etc. These activities are expected to multiply as residents have signified their intent to also undertake commercial activities. With an invigorated business environment, these residents may look forward to a brighter economic future that is not far off from their mainland counterpart.

Environmental Sustainability

The Malalison Island solar PV hybrid pilot project showcased the use of solar PV to mitigate fossil-based generation and expand service availability in an off-grid island, which has resulted in positive socioeconomic development impact. As experienced, the pilot project provided power supply to the island on 24-hour basis, 77% of which came from solar resources. With this, the project was able to mitigate GHG emissions corresponding to approximately 381.48 tons CO_2 equivalent or 84.77 tons CO_2 equivalent annually over the last 4.5 years.

The project overcame the prohibitive cost of power generation using diesel fuel and the logistic challenges that came along with it during monsoon and typhoon seasons. After the project, the island community derived multiple benefits from the availability of sufficient, clean, and reliable electricity for their daily needs 24/7. The residents, especially women and children, were freed from constant exposure to fumes coming from kerosene lamps and indoor pollution from cooking with fuelwood, as electricity for lighting, cooking, and heating appliances became available throughout the day and night. Moreover, the community was able to dispense with the noise pollution coming from several hours of operation of the diesel genset.

The country's carbon footprint in off-grid electrification is high, with 91% of power sourced from fossil fuel. The solar PV hybrid project with battery storage is therefore a technology option that can be adopted in most small islands in the country to reduce dependency on fossil fuel. Aside from its positive environmental impact, the technology has also provided the island some level of resilience over climatic disturbances as well as disruptive events like the COVID-19 pandemic. The operation of the solar PV hybrid project and its prepaid metering feature allowed the people to enjoy the benefits of electricity despite the lockdowns.

Social Welfare Dimension

Electrification is more than just bringing in light. It is also about empowering the community, improving the delivery of basic services, and promoting inclusive growth, ensuring no one is left behind.

From the people's perspective, the project brought numerous positive impacts on the households' living conditions. Foremost is the significant personal and household convenience due to the 24-hour electricity supply the project was able to generate. Women, in particular, appreciated the use of household appliances at any time, which provided convenience and ease of performing in their daily chores. This has opened opportunities for them to generate additional income for the family from activities such as selling ice and frozen foods, cooked meals and/or goods for sale, or extended the operating hours of their small convenience stores. With refrigerators and freezers, fishermen can preserve their catch. Students likewise are able to study at night under better and cleaner lighting conditions as opposed to the use of kerosene lamps. The availability of 24-hour electricity also improved the people's communication among themselves, and with the rest of the country and the world as information are now conveniently available via multimedia such as radio, television, mobile phones, and the internet, among others. Through social media, homestays can advertise and book clients in advance. Entertainment is easily accessed and social functions and community events can be carried out even at night. Taken together, 24-hour electricity service from the operation of the solar PV hybrid project improved the living standards and well-being of the people in Malalison Island.

Second, there was improvement in the basic services provided by the local barangay unit—street lighting for community security, better educational facilities and health services, faster processing of permits, clearances, etc.[18]

Third, the impact of people empowerment is evident in (i) the increased membership and enhanced activities of local organizations, (ii) the ability of the people to adjust to prepaid metering and efficiently manage their power consumption, and (iii) the commerce within the island that prospered in such a short period of time. Even enterprising women, such as Niada Doroteo, at 55 years old, managed to engage in a homestay business thereby earning additional income for herself and her family.

However, as in any intervention, behavioral social changes may occur that could result in negative social impacts. These could affect both families and the community. Surveyed respondents were vocal about some behavioral concerns, which they believe could lead to less time for family and eventual deterioration of relationships within the family and in the barangay. These behavioral concerns include the propensity of people to focus on or pay more attention to electronic gadgets, which could lead to addiction, noisy environment due to socialization until late at night, people engaging in drinking sprees and vices, and family members especially the youth staying up late. Nevertheless, despite these concerns, the substantial and noteworthy positive impact of the project outweighs its perceived disbenefits.

[18] Barangay basic services are made available through direct connection to ANTECO on a postpaid basis as it would be difficult for the barangay office to maintain a prepaid account.

6 POTENTIAL FOR REPLICATION AND FUTURE IMPLICATIONS

Total electrification of the country remains a priority. The Philippine Energy Plan 2023–2040 underscores the goal of total electrification in the country, particularly to reach consumers in off-grid or missionary areas. There are still 6.121 million households yet to be served between July 2023 to 2028. Of these households, 1.285 million are in off-grid areas. Due to the limited or non-access to electricity there is unequal state of development in the country, with many of rural areas' progress hampered by the lack of reliable, affordable, and sufficient electricity supply. Among the major island groups, Mindanao is lagging, with only 86% household electrification rate.

The challenge therefore remains. How to expand the electricity supply and/or provide access to underserved and non-electrified remote communities and off-grid islands in an affordable, reliable, and sustainable manner. Due to the higher cost of generation using imported diesel fuel, many of the islands already served by NPC-SPUG receive only limited hours of electricity. As of December 2022, of the 228 areas that NPC-SPUG serves, only 74 or 32% have 24-hour electricity while 136 or 60% have less than 12 hours of service. Many more have remained without power supply. Renewable energy technologies comprise only 9% of total generation capacity in off-grid areas.

Paul Simon Bertheau, in his doctoral thesis titled "The Role of Renewable Energy for Low-Carbon Development on Small Philippine Islands—A Mixed Methods Approach" concluded that "renewable energy-based island grids represent a cost-effective, environmentally sound, and socially acceptable development pathway for the electricity supply of Philippine islands." Moreover, based on a socioeconomic survey he conducted in Cobrador Island where ADB first supported the implementation of a solar PV hybrid technology, he further concluded that "renewable energy technologies hold substantial potential for improving sustainability and reliability while decreasing costs."[19]

The Malalison Island Solar PV Hybrid Project validates ADB's initial findings in the case of Cobrador Island, that hybridizing the diesel-based power supply generation in small islands in the Philippines is a valid technical solution to facilitate the achievement of the country's off-grid electrification goals. It also affirms that there is private sector interest in off-grid electrification and given the opportunity, the private sector can share and contribute valuable knowledge and management skills to improve and ensure viability of projects. Local governments as well can play a substantive role in making these projects happen in their locality.

[19] P. S. Bertheau. 2020. The Role of Renewable Energy for Low-Carbon Development on Small Philippine Islands—A Mixed Methods Approach. Dissertation. Department of Energy and Environmental Management Europa-Universität Flensburg.

From the Malalison experience, there is substantial evidence showing that the use of solar PV hybrid technology—together with its component parts, i.e., diesel back-up and prepaid metering system—can bring about 24/7 energy access, energy equity, and environmental sustainability to off-grid communities. The increasing trend in energy consumption, the improvement in the basic services by the LGU in terms of health, education, communication, security, climate resilience, and the new income-generating activities initiated as a result of having 24/7 electricity are concrete proofs that it is high time that DRES such as the solar PV hybrid system should be recognized as the pathway to serving off-grid areas in a more sustainable manner.

7 CONCLUSIONS

Clean, sufficient, reliable, and affordable energy supply is critical for reducing poverty and promoting social inclusion and economic growth. As the Government of the Philippines proceeds with its ambitious target of total electrification, some nuggets of wisdom can be derived from the example and experiences of the Malalison Solar PV Hybrid Project.

- There is sufficient evidence that renewable energy can spur growth and welfare benefits in unserved and underserved areas when used to provide energy access and/or enhance and expand service to 24 hours. The use of electricity for income-generating activities and productive uses increases the economic benefits and project sustainability as the market expands.
- Hybridizing diesel-power generation with renewable energy in "missionary electrification" areas is a feasible option to lower the country's carbon footprint and enhance resiliency. Renewable solution, however, need not be solely based on solar PV. Other resources should be considered. Each area has different sets of resource endowment, i.e., could be wind, hydro, biomass aside from solar. Design options, therefore, should follow a differentiated approach. Optimizing the technical design based on the specific resources in the area can be done with the use of optimization tools, such as HOMER, SPT (Simplified Planning Tool, in the case of the Philippines), etc., to ensure that the right set of technologies are considered and optimized.
- Private sector investments can be brought to bear in small island and isolated area electrification. However, it is crucial that government must provide both the policy and market environment to attract these investments. These may include removal of barriers to allow an equal playing field for renewable energy as well as financial incentives, such as partial or smart subsidies to achieve tariff equity for off-grid consumers (for Malalison, partial subsidy came from ADB and the LGU). Government may review current subsidy policies to make them more effective, such as, inclusion of areas served by electric cooperatives in the government's UCME Subsidy Program.
- The introduction of the prepayment system is a plus factor for remote areas. It not only resulted in cost saving and operational efficiency for the electric cooperative—it also allowed inclusivity as electricity can be purchased upon demand and based on the cashflow availability of consumers. Effectively, because consumers can monitor their consumption in real time, they learned to be more judicious in their use of electricity resulting in demand-side energy efficiency.
- Technical and managerial trainings as well as consumer education and participation will help ensure success. Technical know-how, management capability, and local familiarity are critical factors that should be incorporated into project initiatives. Access to appropriate technical and managerial trainings can be made available by the government through its training arm, the Technical Education and Skills Development Authority.

- The transfer of technology from international suppliers must be studied well. In the case of Malalison Island, inadequacy of training from the supplier of the EMS and use of proprietary software gave rise to unnecessary interruptions and prevented the smooth switching of power from solar battery to diesel and vice-versa. A local solution must be developed to resolve this issue. Foreign suppliers currently do not have local after-sales agents because the market is still small to justify investment in this area. Local expertise must be developed to avoid reliance on foreign technology providers.
- Forging of partnerships and collaboration is key to bring in necessary resources into project development and implementation. As in the Malalison story, these were achieved through partnerships between ANTECO, OREEi, LGU, ADB, and the people of Malalison Island. The private sector could provide the technical and management skills. Specifically, in implementing the project, the EC-PSP model worked well for ANTECO. From the initial studies to the project's execution, ANTECO had full support from OREEi, its private sector partner, which provided the expertise in project execution and supported project operations of the power plant and the management of the prepaid metering systems. On financing, development agencies and financing institutions can provide innovative financing tailored to the needs of projects. In the case of pilot project, ADB and NEA, as well as the Land Bank of the Philippines provided innovative financing support. Electric cooperatives can take on the role as agents of change and lead the way to increased DRES deployment in their franchise areas.
- As experienced in many underserved areas, high suppressed demand exists. It is therefore highly possible that in just a short period, supply from the project may soon be outstripped by demand. Thus, in the implementation of mini-grids in underserved areas, project developers should be ready to expand the capacity of the power plant to match increases in demand, otherwise, the energy security in the island may be compromised.
- Since off-grid areas are small, some form of programmatic approach to project development and financing may have to be established for wider deployment. Government should enlist private sector and financial players to map out a workable strategy. Government must also be ready to adjust existing policies and/or come up with new ones to facilitate adoption of innovative solutions.

Through the deployment of DRES, such as the Malalison Solar PV Hybrid Project, it is possible to achieve a balance of energy security, energy equity, and environmental sustainability in remote area electrification. The outcomes of this innovative solution are positive changes in the living conditions of residents and progress in the community. But as demonstrated by the project, to attain real socioeconomic development, 24-hour electricity service is a must have. When this is achieved, local development can begin to happen within a short span of time, such as within 2–4 years after the project's implementation.

Increasing the deployment of DRES would benefit the remaining over 1,200 island-barangays and about 1.29 million households living in off-grid areas without access to sufficient, reliable, and clean energy. Corollary to this, DRES will help achieve the government's goal of attaining 35% renewable energy in the country's energy mix by 2030 and 50% by 2040 or earlier, as embodied in the updated National Renewable Energy Program. Projects like the Malalison Solar PV Hybrid Project should be replicated in other parts of the country and pursued with heightened vigor.

The financing support of development organizations, such as ADB, is a critical ingredient to scale up implementation of projects from pilot phase to full-scale programs. ADB's assistance to ANTECO's solar PV hybrid project is a testament to the bank's commitment to achieving a prosperous, inclusive, resilient, and sustainable Asia and the Pacific, while sustaining its efforts to eradicate extreme poverty and building a sustainable and resilient energy future—ideals that are embodied in ADB Strategy 2030 and 2021 Energy Policy.

The project showcased an innovative solution to the off-grid electrification challenges of the Philippines. Its design is consistent with ADB's focus on innovative technologies and operating business models. It advanced the deployment of renewable energy as an energy access, climate resiliency, and economic solution for remote communities. The introduction of prepaid metering for use in an island setting was valuable as an energy efficiency and inclusivity measure that can be easily deployed. Moreover, the project highlighted the partnerships between public (LGU), private sector (OREEi), ANTECO, and the people of Malalison, ensuring that the project gets full support from all stakeholders. This formula fosters project ownership and assures project sustainability.

Finally, the Malalison Island experience can serve as case study to learn from and hopefully encourage other ADB DMCs to apply renewable energy hybrids technologies, whenever suitable in their local setting.

8 REFERENCES

E. Cecelski. 2003. *Enabling Equitable Access to Rural Electrification: Current Thinking on Energy, Poverty, and Gender.* International Bank for Reconstruction and Development and the World Bank. https://documents1.worldbank.org/curated/en/850681468328564938/pdf/345310Equitable0electrification0access.pdf.

Independent Evaluation Group. 2008. *The Welfare Impact of Rural Electrification: A Reassessment of the Costs and Benefits.* World Bank.

Municipality of Culasi, Antique. https://culasiantique.gov.ph/malalison-island-2/.

J. D. Ocon and P. Bertheau. 2019. Energy Transition from Diesel-based to Solar Photovoltaics-Battery Diesel Hybrid System-based Island Grids in the Philippines – Techno-Economic Potential and Policy Implication on Missionary Electrification. *Journal of Sustainable Development of Energy Water and Environment Systems*. 7(1). pp. 139–154.

www.ingramcontent.com/pod-product-compliance
Lightning Source LLC
LaVergne TN
LVHW070942160826
845679LV00022B/1888
9789292709150